Wind Loads and Anchor Bolts for Petrochemical Facilities

Prepared by the
TASK COMMITTEE ON WIND INDUCED FORCES
and the
TASK COMMITTEE ON ANCHOR BOLT DESIGN of the
PETROCHEMICAL COMMITTEE of the
ENERGY DIVISION of the
AMERICAN SOCIETY OF CIVIL ENGINEERS

Published by
ASCE *American Society of Civil Engineers*
1801 Alexander Bell Drive
Reston, Virginia 20191-4400

Abstract:
Current codes and standards do not address many of the structures found in the petrochemical industry. Therefore, many engineers and companies involved in the industry have independently developed procedures and techniques for handling different engineering issues. This lack of standardization in the industry has led to inconsistent structural reliability. These reports, *Wind Loads on Petrochemical Facilities and Design of Anchor Bolts in Petrochemical Facilities*, are intended as state-of-the-practice set of guidelines in the determination of wind induced forces and the design of anchor bolts for petrochemical facilities, respectively. These reports are aimed at structural design engineers familiar with design of industrial-type structures.

Library of Congress Cataloging-in-Publication Data

Wind loads and anchor bolt design for petrochemical facilities / prepared by the Task Committee on Wind Induced Forces and the Task Committee on Anchor Bolt Design of the Petrochemical Committee of the Energy Division of the American Society of Civil Engineers.
  p.   cm.
  ISBN 0-7844-0262-0
  1. Petroleum refineries--Design and construction. 2. Wind-pressure. I. American Society of Civil Engineers. Task Committee on Wind Induced Forces. II. American Society of Civil Engineers. Task Committee on Anchor Bolt Design
TH4571.W55 1997   97-20890
693.8'5--dc21   CIP

     The material presented in this publication has been prepared in accordance with generally recognized engineering principles and practices, and is for general information only. This information should not be used without first securing competent advice with respect to its suitability for any general or specific application.
     The contents of this publication are not intended to be and should not be construed to be a standard of the American Society of Civil Engineers (ASCE) and are not intended for use as a reference in purchase specifications, contracts, regulations, statutes, or any other legal document.
     No reference made in this publication to any specific method, product, process or service constitutes or implies an endorsement, recommendation, or warranty thereof by ASCE.
     ASCE makes no representation or warranty of any kind, whether express or implied, concerning the accuracy, completeness, suitability, or utility of any information, apparatus, product, or process discussed in this publication, and assumes no liability therefore.
     Anyone utilizing this information assumes all liability arising from such use, including but not limited to infringement of any patent or patents.

Photocopies. Authorization to photocopy material for internal or personal use under circumstances not falling within the fair use provisions of the Copyright Act is granted by ASCE to libraries and other users registered with the Copyright Clearance Center (CCC) Transactional Reporting Service, provided that the base fee of $4.00 per article plus $.25 per page is paid directly to CCC, 222 Rosewood Drive, Danvers, MA 01923. The identification for ASCE Books is 0-7844-0262-0/97/$4.00 + $.25 per page. Requests for special permission or bulk copying should be addressed to Permissions & Copyright Dept., ASCE.

Copyright © 1997 by the American Society of Civil Engineers,
All Rights Reserved.
Library of Congress Catalog Card No: 97-20890
ISBN 0-7844-0262-0
Manufactured in the United States of America.

Wind Loads on Petrochemical Facilities

Prepared by the

Task Committee on Wind Induced Forces

The ASCE Petrochemical Energy Committee

This publication is one of five state-of-the-practice engineering reports produced, to date, by the ASCE Petrochemical Energy Committee. These engineering reports are intended to be a summary of the current knowledge and design practice, and present guidelines for the design of petrochemical facilities. They represent a consensus opinion of task committee members active in their development. These five ASCE engineering reports are:

1) *Design of Anchor Bolts in Petrochemical Facilities*

2) *Design of Blast Resistant Buildings in Petrochemical Facilities*

3) *Design of Secondary Containment in Petrochemical Facilities*

4) *Guidelines for Seismic Evaluation and Design of Petrochemical Facilities*

5) *Wind Loads on Petrochemical Facilities*

The ASCE Petrochemical Energy Committee was organized by A. K. Gupta in 1991 and initially chaired by Curley Turner. Under their leadership, the task committees were formed. More recently, the Committee has been chaired by J. A. Bohinsky followed by Frank Hsiu.

Frank Hsiu
Chevron Research and Technology Company
chairman

J. Marcell Hunt
Hudson Engineering Corporation
secretary

Joseph A. Bohinsky — Brown & Root, Inc.
William Bounds — Fluor Daniel, Inc.
Clay Flint — Bechtel, Inc.
John Geigel — Exxon Chemical Company
Ajaya K. Gupta — North Carolina State University
Magdy H. Hanna — Jacobs Engineering Group
Steven R. Hemler — Eastman Chemical Co.
Gayle S. Johnson — EQE International, Inc.
James A. Maple — J. A. Maple & Associates
Douglas J. Nyman — D. J. Nyman & Associates
Norman C. Rennalls — BASF Corporation
Curley Turner — Fluor Daniel, Inc.

The ASCE Task Committee on Wind Induced Forces

This report is intended to be a state-of-the-practice set of guidelines. It is based on reviews of current practice, internal company standards, published documents, and the work of related organizations. The report includes a list of references that provide additional information.

This report was prepared to provide guidance in the determination of wind induced forces for petrochemical facilities. However, it should be of interest to structural design engineers familiar with design of industrial type structures and the application of *ASCE 7 "Minimum Design Loads for Buildings and other Structures"* to these type structures.

The committee would like to thank Ahmad Nadeem who was assisted greatly with our research on open frame structures.

Norman C. Rennalls	Jon Ferguson
BASF	Brown & Root, Inc.
chairman	secretary

Nguyen Ai	Jacobs Engineering Group
John Geigel	Exxon Chemical Company
Udaykumar Hate	M. W. Kellogg Company
Manuel Heredia	John Brown Engineers
Marc Levitan	Louisiana State University
Marvin Lisnitzer	Stone & Webster
James Maple	J. A. Maple & Associates
Pravin Patel	E.I. Du Pont De Nemours and Company
Ted Puteepotjanart	Raytheon Engineers and Constructors
Ashvin Shah	Fluor Daniel, Inc.
Jerry Suderman	Bechtel, Inc.
John Tushek	Dow Chemical

Reviewers

Al Wussler	El Paso Natural Gas / Gas Processors Association
David Kernion	RPM Engineering Inc.
Jon Peterka	Cermak Peterka Petersen, Inc.

CONTENTS

Chapter 1: Introduction 1-1

    1.1   Background 1-1
    1.2   State of the Practice 1-2
    1.3   Purpose of Report 1-2

Chapter 2: Survey of Current Design Practices 2-1

    2.1   Introduction 2-1
    2.2   Pipe Racks 2-1
    2.3   Open Frame Structures 2-4
    2.4   Pressure Vessels 2-6

Chapter 3: Comparisons of Design Practices 3-1

    3.1   Introduction 3-1
    3.2   Pipe Racks 3-2
    3.3   Open Frame Structures 3-5
    3.4   Pressure Vessels 3-11

Chapter 4: Recommended Guidelines 4-1

    4.0   General 4-1
    4.1   Pipe Racks 4-2
    4.2   Open Frame Structures 4-3
    4.3   Pressure Vessels 4-16
    Appendix 4A Alternate Method for Determining
              C_f and Load Combinations for Open Frame Structures 4-21

Chapter 5: Examples 5-1

    Appendix 5A   Example - Pipe Racks 5-3
    Appendix 5B   Example - Open Frame Structures 5-9
    Appendix 5C   Example - Pressure Vessels 5-19

Chapter 6: Research Needs 6-1

    6.0   General 6-1
    6.1   Research Priorities 6-2

CONTENTS (Cont'd)

Nomenclature A-1
Glossary B-1
References C-1

CHAPTER 1
INTRODUCTION

This report is structured around generic types of facilities usually found in the process industries:

   a) Pipe support structures (pipe racks).
   b) Open and partially clad frame structures.
   c) Vessels (vertical, horizontal and spherical).

1.1 BACKGROUND

The basis and procedures for determining wind induced forces for enclosed structures and other conventional structures are well documented in the engineering literature. These design basis and procedures have been adopted by ASCE and codified in *ASCE 7* and its predecessor documents. Other organizations have incorporated the major provisions of *ASCE 7* into building codes, including the Uniform Building Code, Standard Building Code and BOCA/National Building Code. These building codes have been adopted in ordinances and laws written by various local and regional jurisdictions.

The "Scope" statement for *ASCE 7* indicates that the standard provides minimum load requirements for the design of buildings and other structures that are subject to building codes. *ASCE 7* does not adequately address open frame structures, structures with interconnecting piping, partially clad structures, and vessels with attached piping and platforms. However, it does address enclosed structures, trussed towers and simple cylinders.

Wind induced forces are typically calculated using the force equation from *ASCE 7*:

$$F = q_z \, G \, C_f \, A \qquad (1.1)$$

In this equation q_z is the velocity pressure component, G is for the gust component, C_f is the force/shape/drag/shielding component, and A is the area for which the force is calculated. The velocity pressure component of this force (q_z) has three factors; the importance of the structure, the surrounding terrain (exposure category), and design wind speed.

The selection of basic wind speed, importance factor, exposure category and gust response factor are defined in *ASCE 7* and therefore are not discussed in detail. Force coefficients, tributary areas, and shielding are not clearly defined in *ASCE 7* for industrial - type structures and equipment. These load components are discussed in this report and recommendations for selecting values are made. Since this report is intended to supplement *ASCE 7*, the designer will be referred to that document when it provides the appropriate information. The nomenclature and glossary used in the recommendations of this document mirror those found in *ASCE 7*.

1.2 STATE OF THE PRACTICE

This study is based on current industry practices in the design of petrochemical facilities. The practices are generally based on a company's experience and the desire to provide an economical facility that provides a margin of safety that is consistent with the perceived risk. These practices, as interpreted by the committee, are quite varied. For a given type of structure, the practices currently in use can result in design wind induced forces that vary by factors as large as 5, when using the same basic wind speed and exposure category.

1.3 PURPOSE OF REPORT

It is intent of this committee that the publication of this report will result in a more uniform application of practices across the petrochemical energy industry. In order to facilitate this goal a set of recommended guidelines is presented as part of this report.

CHAPTER 2
SURVEY OF CURRENT DESIGN PRACTICES

2.1 INTRODUCTION

Thirteen design practices, for wind on pipe racks, open structures and pressure vessels were reviewed. These design practices were obtained from various operating companies and engineering contractors working in the petrochemical industry.

All but one of the design practices reviewed were based on the wind loading provisions of *ASCE 7*-88, *"Minimum Design Loads for Buildings and Other Structures"* or its predecessor (*ANSI A58.1*). The recommended wind speed, mean recurrence interval and exposure coefficient (based on terrain exposure C, open terrain) were generally the same.

The building classification, used for the importance factor, by all of the design practices was ordinary structures (Category I). However, a few practices chose to use the essential facilities category (Category III) for particular structures.

2.2 PIPE RACKS

Most of the design practices treated the rack structure as an open frame structure with additional loads for pipe and cable trays. Since open frame structures are discussed in another section of this report the wind loads applied directly on the structure will not be discussed here. Table 2.1 presents the survey results in tabular form.

The major differences between design practices is covered in Table 2.1. Note that the determination of force coefficient, effective area, and shielding is not addressed in *ASCE 7* for piping and cable trays in a pipe rack therefore the following definitions are used Table 2.1 as defined by the respective design practice.

A Tributary area (sq. ft).
d Depth of cable tray (ft).

TABLE 2.1
Survey Of Piperack Design Practices

DESIGN GUIDE	BASIS CODE	FORCE COEFFICIENT	TRIBUTARY AREA	SHIELDING/COMMENTS
#1	ASCE 7-88	Part I $C_f = 0.7 + W/25D < 2$ for pipes $C_f = 2.0 + W/25d < 4$ for cable tray Part II $C_f = 1.3$	Part I $A = D*L$ for pipes $A = d*L$ for cable tray Part II $A = H*L$	Shielding is taken care of in the Force Coefficient. Part III Use the smaller value from Part I or II.
#2	ASCE 7-88	$C_f = 0.8$ (Includes the Gust Factor) for pipes.	$A = D_{avg}*m*L$ for pipes $A = d*m*L$ for cable tray	Shielding is included in the factor m.
#3	ASCE 7-88	$C_f = 1.1 * \tan 10° = 0.19$ for pipe and cable tray or $C_f = 0.8$	$A = W*L$ or $A = D*L$	Shielding is included in C_f and A. The force is based on the wind at a 10° angle of incidence and a force coefficient of 1.1. Force shall not be less than on the largest pipe alone.
#4	ANSI A58.1 '82	Not specified.	Not specified.	Does not address pipe or cable tray.
#5	ASCE 7-88	Not specified. Engineer shall interpret ASCE. Alternatively $C_f = 0.31$ for $W \geq 4$ ft (1.22 m) $= 0.22$ for $W \geq 16$ ft (4.88 m)	Area is based on an angle of incidence of ±10° and a 5 dia. shielding cone. $A = (D+(W-D)\sin 15.2°)\cos 10° L$ $= D+0.25(W-D) L$ Alternatively $A = W*L$	Shielding is included in area. Alternately Shielding is included in force coefficient and area. Does not address cable tray.
#6	ASCE 7-88	$C_f = 0.7$ for $D\sqrt{q_z} > 2.5$ $= 1.2$ for $D\sqrt{q_z} \leq 2.5$	$A = [1+W/6D]*D*L$	Wind is based on an angle of incidence of 10°. Does not address cable tray.

TABLE 2.1
Survey Of Piperack Design Practices (cont'd)

DESIGN GUIDE	BASIS CODE	FORCE COEFFICIENT	TRIBUTARY AREA	SHIELDING/COMMENTS
#7	ASCE 7-88	$C_f = 0.7$ for pipes $C_f = 1.75$ for cable trays	$A = D_{avg} \cdot SM \cdot L$ for pipe $A = d \cdot (1+(N-1)m) \cdot L$ for cable trays	Shielding is incorporated in factor SM for pipes and n for cable trays.
#8	ASCE 7-88	Not specified.	Not specified.	Uses Exposure C outside battery limits and B inside.
#9	ASCE 7-88	From Table 14 of ASCE 7-88 For pipes use $C_f = D\sqrt{q_z} > 2.5$ For cable tray use C_f for flat surfaces.	$A = (D+10\%$ of usable width for pipe or cable tray$) \cdot L$.	Shielding is included in area. Uses Exposure C or B.
#10	ANSI A58.1 1982	$C_f = 1.0$ when used for all components or use C_f values from ANSI when applied to individual components.	$A = 3 \cdot L$ for racks > 12 ft (3.66 m) $A = 2 \cdot L$ for racks < 12 ft (3.66 m) D > 16 inches need special consideration.	Shielding is included in area. Does not address cable tray.
#11	ASCE 7-88	$C_f = 0.7$	$A = 3 \cdot L$ for racks W< 12 ft (3.66 m) $A = 2 \cdot L$ for racks 12<W<20 ft (3.66 - 6.10 m)	Shielding is included in area. Does not address cable tray.
#12	ASCE 7-88	Not specified.	Not specified.	
#13	UBC or ANSI A58.1 1982	$C_f = 2$	$A = D \cdot L$	Shielding is included in force coefficient. Does not address cable tray.

D_{avg} Average diameter of all pipes greater than 6 in. (0.15 m) plus the average diameter of the four largest pipes (ft).
D   Diameter of the largest pipe (ft).
H   Distance from the bottom level to the top level of the pipe rack (ft).
L   Pipe rack bent spacing (ft).
m   A factor to account for shielding, varying from 1.0 for one pipe to 3.3 for 12 pipes or more.
n   A shielding factor for cable trays varying depending on the ratio of tray height to spacing between the adjacent trays. Ranging from 0 to 0.5.
N   Number of cable trays.
q_z Velocity Pressure (psf)
SM   A factor to account for shielding, varying from 1.0 for one pipe to 4.5 for 12 pipes or more.
W   Width of pipe rack (ft).

2.3 OPEN FRAME STRUCTURES

The design practices divide their evaluation of open frame structures into three general areas:

- Equipment
- Structural framing
- Piping & misc. attachments (ladders, handrail etc.)

Generally, the wind load on the structure and equipment supported by the structure are considered separately. No shielding effect between equipment or between structure and equipment is considered. However, several practices place an upper bound on the total wind load on the structure and therefore indirectly reflect shielding.

Wind loads on equipment are computed either from principles of *ASCE 7* or as described in the following section on pressure vessels. Additionally, other than what was discussed in the previous section on pipe racks the differences between how the different practices handled piping and miscellaneous attachments in open frame structures was minor. Therefore, this section concentrates on the differences on how the structural framing is evaluated in the various design practices. Table 2.2 presents the survey results.

In the case of flexible structures, the procedure given in the commentary section C6.6 of *ASCE 7*-88 is generally, with one exception, recommended by each standard.

TABLE 2.2
Survey of Design Practices for Open Frame Structures

Design practice	Basis	Comments
1	Whitbread & ASCE 7	Whitbread is the basis for shielding factors. For six frames or less an enclosed structure is upper bound. There is provision for increasing the force coefficient for more than 6 frames. Considers load out of horizontal plane. The area considered is the projected area of the windward frame.
2	ASCE 7 & National Building Code of Canada	Force coefficient is based on National Building Code of Canada
3	ASCE 7 & Georgio/Vickery/Church	Force coefficient is based on Georgio/Vickery/Church. A single C_f factor is considered in the calculation of wind on the structure C_f = 1.8 for all the framing members and 1.6 for ladders and platforms.
4	ASCE 7	Provides simplified combined load coefficient for the gross area of the structure.
5/6	Whitbread & ASCE 7	Whitbread is the basis for shielding factors. The total wind force is limited to the upper bound of wind load on an enclosed structure that would completely envelop the structure and attachments.
7	ASCE 7	The calculation of wind load considers the projected area of: a) the deepest girder for a floor level b) depth of flooring c) piping and electrical as a percentage of the gross area d) handrails perpendicular and parallel to the direction of wind, e) stairs f) projected area of all columns and all vertical bracing (no shielding)
8	ASCE 7	Use Exposure B for pipe racks. No other guidelines provided.
9	ASCE 7 & AS1170, Part 2, Standards Association of Australia	C_f is taken from Table 14 and 15 of ASCE 7-88. Shielding is based on AS1170, Part 2, Standards Association of Australia
10/11	ASCE 7	Proprietary, simplified "effective" force coefficient applied to the gross area of the structure.
12	ASCE 7	C_f is based on Table 14 of ASCE 7-88. Proprietary force coefficient applied to the windward frame.
13	ANSI A58.1	The force coefficients used for open frame structures and pipe racks are 1.3 for the windward frame, 0.8 for the 2nd frame, and 0.5 for the 3rd and succeeding frames.

2.4 PRESSURE VESSELS

The wind load determinations for vertical and horizontal pressure vessels for all design standards are predominantly in accord with *ASCE 7*-88, however there are some key differences in the approaches taken.

The main differences involve the effects of platforms, piping, ladders, nozzles and insulation. These differences have the most effect on total shear and moment for horizontal vessels. For platforms, some of the guidelines use plan projections of platforms, some use percentage of vessel diameter, some use vertical projected area, and some use projection of individual members. Most of the guidelines add a percentage of vessel diameter or 1 to 2 ft (0.30 to 0.61 m) to the diameter to account for appurtenances such as insulation, nozzles, and ladders. For some guidelines large pipe is accounted for separately and for others a percentage of the vessel diameter is used. Force coefficients for pipe vary from about 0.6 to 1.6 (most being about 0.7). The coefficients for the vessel itself are generally in line with *ASCE 7*-88, Table 12, but values for various surface roughness differ.

Most design practices state that if a vessel has an unusual amount of piping, platforms, etc. or is of an unusual configuration, the components should be figured separately instead of lumping them with a simplifying factor.

Most design practices require calculation of a gust response factor for flexible structures if the h/D ratio exceeds 5 or the fundamental frequency is less that 1 Hz. Table 2.3 presents the survey results in tabular form.

TABLE 2.3

Survey of Design Practices For Vessels

Design Practice	Notes
1.	**Vertical Vessels** - A_f based on increased shell diameter to account for manways, ladders, platforms and small piping. C_f is for moderately smooth cylinders, but h/D break points are at 4 and 16 in lieu of 7 and 25 (as in ASCE 7 -88 Table 12). **Horizontal Vessels** - A_f based on increased shell diameter to account for supports, piping and other attachments. C_f is 0.5 or 0.6 depending on slenderness ratio for transverse wind; and, for longitudinal wind, 1.0 for flat heads and 0.5 for rounded heads.
2.	**Vertical Vessels** - A_f is based on actual shell diameter. C_f is the larger of that for rough surfaces or that for moderately smooth surfaces multiplied by a factor based on vessel diameter, to cover ladders and piping. Platforms are figured separately and A_f is based on 1/2 the platform plan area. **Horizontal Vessels** - Platforms figured separately and A_f is based on 1/2 the platform plan area.
3.	**Vertical Vessels** - A_f for h<75 ft.(22.86 m) based on vessel diameter multiplied by a factor depending upon vessel diameter to cover appurtenances. C_f is for rough surfaces. A_f for h>75 ft. (22.86 m) is based on actual vessel diameter. C_f is for rough surfaces. Platforms, ladders and piping figured separately. **Horizontal Vessels** - A_f is based on increased vessel diameter to cover manways, platforms and piping.
4.	No simplifying assumptions presented for vessels.
5.	**Vertical Vessels** - A_f is actual vessel diameter. C_f for very rough cylinders is applied over a portion (not less than 50%) of the vessel diameter to account for ladders, platforms and piping with diameters less than 5% of the vessel diameter. C_f for moderately smooth cylinders is applied to the remaining shell diameter. Area of piping with diameters greater than 5% of the vessel diameter is added and C_f is 1.4. **Horizontal Vessels** - No simplifying assumptions presented.
6.	**Vertical Vessels** - A_f is actual vessel diameter. C_f for very rough cylinders is applied over a portion (not less than 50%) of the vessel diameter to account for ladders, platforms and piping with diameters less than 5% of the vessel diameter. C_f for moderately smooth cylinders is applied to the remaining shell diameter. Area for piping with diameters greater than 5% of the vessel diameter is added and C_f is 1.4. **Horizontal Vessels** - No simplifying assumptions presented.

TABLE 2.3 (Cont'd)

Survey of Design Practices For Vessels

Design Practice	Notes
7.	No simplifying assumptions presented for vessels.
8.	No simplifying assumptions presented for vessels.
9.	**Vertical Vessels** - A_f based on vessel diameter multiplied by a factor depending upon vessel diameter to account for nozzles, manways, piping and insulation. C_f is for moderately smooth cylinders. **Horizontal Vessels** - Same approach as for vertical vessels is used.
10.	**Vertical Vessels** - A_f is based on vessel diameter plus largest pipe plus 1.0 ft. (0.3 m) to cover ladders and small piping. C_f is for moderately smooth cylinders. Platforms are figured separately and A_f is based on 1/2 the platform plan area. **Horizontal Vessels** - Same approach as for vertical vessels is used.
11.	**Vertical Vessels** - A_f is based on vessel diameter plus largest pipe plus 1.0 ft. (0.3 m) to cover ladders and small piping. C_f is for moderately smooth cylinders. Platforms are figured separately and A_f is based on 1/2 the platform plan area. **Horizontal Vessels** - Same approach as for vertical vessels is used.
12.	**Vertical Vessels** - A_f is based on vessel diameter multiplied by 1.2 to account for piping and other appurtenances. **Horizontal Vessels** - A_f is same as for vertical vessels. C_f is 0.8.
13.	**Vertical Vessels** - A_f is based on the vessel diameter plus the largest pipe plus a factored platform width. **Horizontal Vessels** - A_f is the vessel diameter multiplied by a factor dependent on vessel diameter.

CHAPTER 3
COMPARISONS OF DESIGN PRACTICES

3.1 INTRODUCTION

Since shielding is sometimes included in the force coefficient and other times in the projected area, the design practices cannot be compared by simply looking at the various components of wind load (i.e. force coefficient, projected area, and shielding). Therefore, typical examples of structures encountered in practice have been selected. These examples are used to create tables that compare the applied wind forces for each design practice. The tables also include the wind forces resulting from the methods proposed in Chapter 4 used with *ASCE* 7-88, to match the design practices, and the proposed methods used with the current standard, *ASCE* 7-95.

The definitions and values for structure classification categories and corresponding importance factors as well as the wind speed map have changed in the newer version of the standard.

- Using the definitions of *ASCE* 7-88 structures were assumed to be classified as Category I ('ordinary') structures (*ASCE* 7-88, Table 1). In *ASCE* 7-95, there is the possibility of choosing between Category II and Category III for the wind induced forces. Category II is now 'ordinary' structures, what used to be called Category I in *ASCE* 7-88. Category III includes "...structures that represent a substantial hazard to human life in the event of failure."

- In *ASCE* 7-95, the wind speed map has been changed from fastest mile to a three-second design wind velocity. There were also corresponding changes made in the velocity pressure exposure coefficients and gust factors.

The structures used in the comparisons were set in a fictitious plant in an arbitrary location. The location chosen was Lake Charles, Louisiana (in southwest Louisiana, about 20 miles from the gulf coast). The wind speed and importance factor used in the comparisons were 96 mph (43 m/s) and 1.04 respectively (*ASCE* 7-88, Figure 1 and Table 5). Terrain exposure C was assumed (flat, open terrain with scattered obstructions).

Wind loads calculated using the proposed guidelines of Chapter 4 with *ASCE* 7-88 and *ASCE* 7-95 with the 'ordinary' structure classification differed only slightly. The differences are due to the change in the wind speed definition and the vertical velocity profile between the standards. Wind loads using *ASCE* 7-95 and Category III structures were 15% greater than those for an 'ordinary' structure classification, due to increased importance factor.

3.2 PIPE RACKS

The following pipe and cable tray configurations (Pipe Load Cases) were used for comparison.

- Case I 20 ft (6.10 m) wide rack with one 48 in (1.22 m) pipe & fourteen 9 in (0.23 m) pipes
- Case II 20 ft (6.10 m) wide rack with fifteen 12 in (0.30 m) pipes
- Case III 5 ft 6 in (1.68 m) wide rack with one 24 in (0.61 m) pipe & two 12 in (0.30 m) pipes
- Case IV 4 ft 6 in (1.37 m) wide rack with three 12 in (0.30 m) pipes
- Case V 20 ft (6.10 m) wide rack with one 36 in (0.91 m) pipe, two 24 in (0.61 m) pipes, four 12 in (0.30 m) pipes, & six 9 in (0.23 m) pipes
- Cable Tray 20 ft (6.10 m) wide rack with two 36 in (0.91 m) trays, one 24 in (0.61 m) tray, two 18 in (0.46 m) trays, two 12 in (0.30 m) trays, & two 6 in (0.15 m) trays, all 6 in (0.15 m) high.

These pipe load cases are illustrated in Figure 3.1. Table 3.1 summarizes the wind forces calculated using the design practices. In order to have a common basis for comparison, the following criteria were used:

- Height of the pipe or cable tray level used for calculations: 30 ft (9.14 m)
- Velocity Pressure Exposure Coefficient (for z = 30 ft (9.14 m)): K_z = 0.98
- Velocity Pressure : $q_z = 0.00256\ K_z(IV)^2 = 25.1\ psf\ (1.2\ kN/m^2)$
- Gust Response Factor : G_h = 1.26
- Design Wind Pressure : $q_z\ G_h = 31.6\ psf\ (1.51\ kN/m^2)$

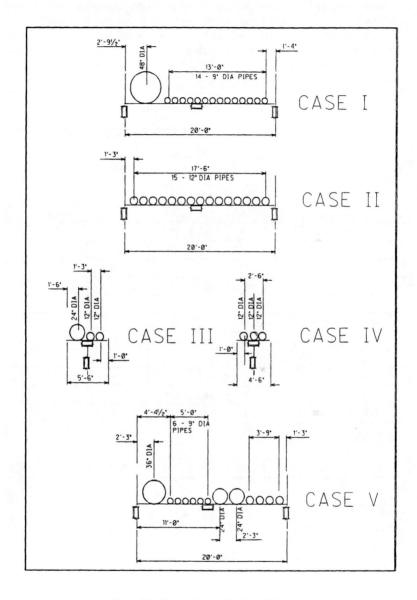

Figure 3.1 Comparison - Pipe Load Cases

This section is limited to wind force on the piping or cable trays only; therefore, wind on the structure has not been included. In addition to the design practices, load cases A and B are included to establish lower and upper bound values for wind forces on pipes. Basis A is wind on the largest pipe only and can be thought of as a minimum probable load (lower bound). Basis B is full wind on all the pipes (no shielding) and can be thought of as the maximum probable load (upper bound).

TABLE 3.1 Design Practice Comparisons - Pipe Rack Wind Forces

VALUES ARE IN POUNDS PER LINEAR FOOT OF PIPE OR CABLE TRAY (Note 1)							
BASIS	PIPE LOAD CASES					CABLE TRAYS	REMARKS
	I	II	III	IV	V		
A	66	22	44	22	50	-	Lower bound
B	299	332	88	66	343	-	Upper bound
#1	114	48	51	28	92	57	
#2	64	66	59	44	104	(Note 3)	
#3	121	121	51	27	121	121	
#4	-	-	-	-	-	-	(Note 2)
#5	167	127	66	41	167	(Note 3)	
#6	167	127	66	41	167	(Note 3)	
#7	102	100	65	49	156	61	
#8	-	-	-	-	-	-	(Note 2)
#9	202	97	70	34	169	121	
#10	179	95	84	63	190	(Note 3)	
#11	93	93	62	62	93	(Note 3)	
#12	-	-	-	-	-	-	(Note 2)
#13	253	63	126	63	190	(Note 3)	

TABLE 3.1 Design Practice Comparisons - Pipe Rack Wind Forces (Cont'd)

BASIS	VALUES ARE IN POUNDS PER LINEAR FOOT OF PIPE OR CABLE TRAY (Note 1)						REMARKS
	PIPE LOAD CASES					CABLE TRAYS	
	I	II	III	IV	V		
Avg. Practices #1-#13	146	94	70	45	145	90	
Recommended Design Practice (See Chap 4)	133	66	56	32	111	158	ASCE 7-88
	129	64	55	31	107	153	ASCE 7-95 Category II
	148	74	63	36	123	176	ASCE 7-95 Category III

Notes to Table 3.1

1 To convert pounds per foot to newtons per meter multiply values by 14.6.

2 Unable to calculate load values from the provided documents.

3 This design practice does not address cable trays.

3.3 OPEN FRAME STRUCTURES

The plan and elevation views of the structure used for comparison are shown in Figures 3.2, 3.3 and 3.4. The structure considered was 40 ft (12.19 m) x 40 ft (12.19 m) x 82 ft (24.99 m) high, with three open frames in the direction of wind.

The structure supported two horizontal vessels (4 ft (1.22 m) diameter x 10 ft (3.05 m) long & 16 ft (4.88 m) diameter x 32 ft (9.75 m) long) on level 20 ft 0 in (6.10 m) and three horizontal exchangers (2 @ 10 ft (3.05 m) diameter x 24 ft (7.32 m) long & 2 ft (0.61 m) diameter x 20 ft (6.10 m) long) on level 48 ft 0 in (14.63 m).

Member sizes were assumed as follows:

    Columns   -   12 in x 12 in (0.31 m x 0.31 m)
    Beams El 20 ft 0 in (6.10 m) -   W36

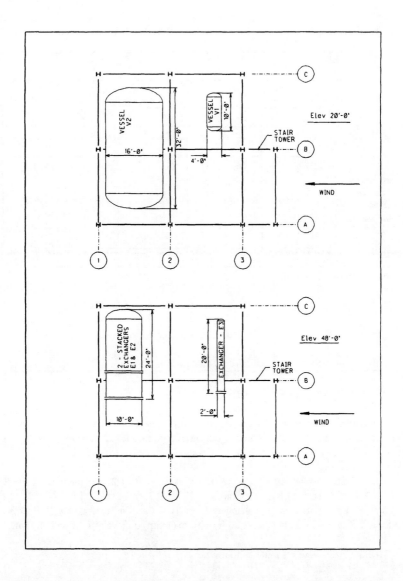

Figure 3.2 Example for Open Frame Structure Comparison - Arrangement Plan

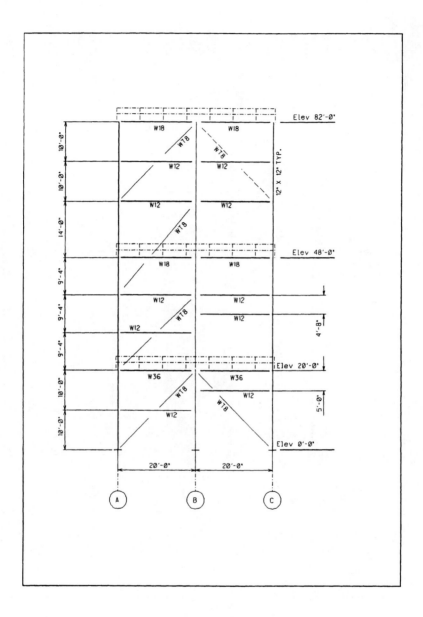

Figure 3.3 Example for Open Frame Structure Comparison - Column Line 3

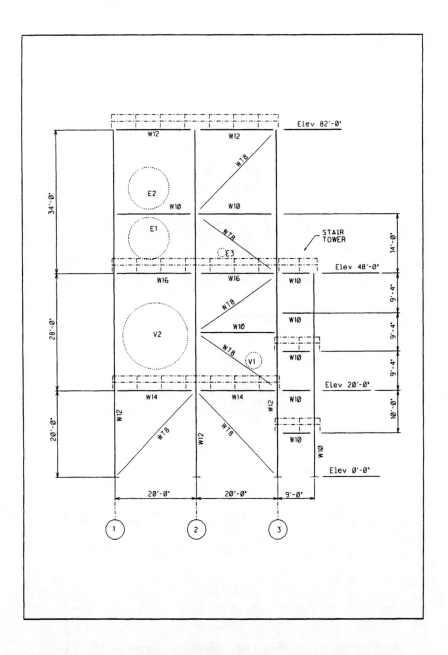

Figure 3.4 Example for Open Frame Structure Comparison - Column Line A

Beams El 48 ft 0 in (14.63 m) -   W18
Beams El 82 ft 0 in (24.99 m) -   W18
Braces   -   W8
Intermediate Beams   -   W12

For this comparison the windward frame was column line 3. The projected areas of the stair tower members were included in the calculation of the windward frame (see example section 5B.1 for details).

The calculations considered provisions for the projected area for piping and electrical trays per the design practices. If these provisions were not defined in the design practice the reviewers chose a projected area of piping and electrical to be 20% of the equipment area.

The results of total wind force are tabulated in Table 3.2. The comparison of the results of the calculations categorizes the design practices in three groups.

1. Some companies use C_f factor based on Table 15 of *ASCE 7* and consider only first two frames, resulting in lower wind force.

2. Some companies limit total wind force by an upper bound based on a totally enclosed structure. The results of total wind force are almost the same.

3. Some companies consider multiframe open structures with or without shielding. No upper bound is considered. Generally the C_f factor considered for the open frame structure remained constant for these practices. However, the areas exposed to the wind differed and hence the results.

TABLE 3.2 Design Practice Comparisons - Open Frame Structure Wind Force

DESIGN PRACTICE	VALUES ARE BASE SHEAR IN KIPS (Note 1)		COMMENTS
	COMBINATION A (Note 2)	COMBINATION B (Note 3)	
#1	NA	162	See Note 4.
#2	122	160	Open frame structure considered.
#3	118	148	Open frame structure considered.
#4	88	112	Open frame structure considered.
#5	130	141	See Note 4.
#6	130	141	Same as Design Practice #5.
#7	134	154	Open frame structure considered.
#8	105	129	Two frames considered per Table 15 of ANSI.
#9	97	120	See Note 4.
#10	93	116	Has used an effective force coefficient other than from ASCE or ANSI.
#11	NA	90	Effective force coefficient based on the solidity ratio of the structure.
#12	157	182	Similar frames assumed in an open frame structure.
#13	92	116	Open frame structure considered.
Average	115	136	
Recommended See Chap 4	123	147**	*ASCE 7-88*
	124	148**	*ASCE 7-95* Category II
	143	170**	*ASCE 7-95* Category III

Notes to Table 3.2

1. To convert kips to kilonewtons multiply values by 4.448.
2. Combination A includes only wind on the structural frame. When NA is reported the design practice did not segregate the results between equipment and the structure in this case.
3. Combination B includes wind on the structural frame, equipment, piping, and electrical.
4. Limited by the upper bound which is typically an enclosed structure.

** Must be applied simultaneously with partial wind load on the other structural axis. See section 4.2.6 and Appendix 4A.

3.4 PRESSURE VESSELS

3.4.1 Vertical Vessels

In order to have a common basis for comparison, a set of criteria and vessel configurations were developed. The example vertical vessel along with its criteria is shown on Figure 3.5. Calculations of wind forces from contributing components were prepared and combined into loads at the base of the vessel. Table 3.3 presents a numerical comparisons.

Forces for rigid and flexible structures (fundamental natural frequency greater than or less than 1 Hz, respectively) were calculated for each design practice. Recommendations for dynamic analysis due to vortex shedding, flutter, etc. is beyond the scope of this report.

For the selected tower the empty natural frequency was 1.18 Hz, more than 1.0, and per *ASCE 7* a gust factor for rigid structures is therefore used to obtain wind forces. When the operating weight of the tower was used, the natural frequency was 0.91 Hz. Since this was less than 1.0, it was necessary to calculate a gust factor for flexible structures ($\overline{G}$). Methods in the design practices were used, when available, to calculate $\overline{G}$. If the design practice did not provide a method to calculate the gust factor, the method in *ASCE 7* commentary was used.

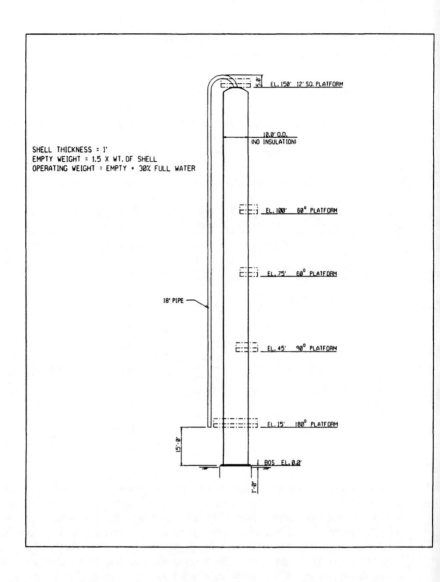

Figure 3.5 Example for Vertical Vessel Comparison

Table 3.3 Design Practice Comparison - Vertical Vessel Wind Load

DESIGN PRACTICE NO.	VALUES ARE BASE SHEAR IN KIPS (Note 1)							
	BASED ON A RIGID VESSEL				BASED ON A FLEXIBLE VESSEL			
	TOTAL SHEAR	PLATFORM SHEAR	VESSEL+ MISC. SHEAR	PIPE SHEAR	TOTAL SHEAR	PLATFORM SHEAR	VESSEL+ MISC. SHEAR	PIPE SHEAR
#1	49.3	3.3 (Note 3)	45.3	0.8 (Note 5)	62.4	4.2 (Note 3)	57.2	1.0 (Note 5)
#2	50.4	5.3	45.2	(Note 4)	63.9	6.7	57.2	(Note 4)
#3	65.3	4.6	53.9	6.9	79.7	5.6	65.7	8.4
#4	(Note 2)							
#5	61.9	12.5	36.3	13.2	77.1	15.6	45.2	16.4
#6	61.9	12.5	36.3	13.2	77.1	15.6	45.2	16.4
#7	(Note 2)							
#8	(Note 2)							
#9	55.0	9.9	40.0	5.1	68.8	12.4	50.0	6.4
#10	50.1	5.4	39.3	5.4	59.8	6.6	46.8	6.4
#11	50.1	5.4	39.3	5.4	59.8	6.6	46.8	6.4
#12	53.5	12.4	34.3	6.9	66.4	15.3	42.5	8.5
#13	48.7	3.9	38.6	6.3	55.5	4.4	44.0	7.2

Table 3.3 Design Practice Comparison - Vertical Vessel Wind Load (Cont'd)

VALUES ARE BASE SHEAR IN KIPS
(Note 1)

AVERAGE & RECOMMENDED PRACTICE (SEE CHAPTER 4)	BASED ON A RIGID VESSEL				BASED ON A FLEXIBLE VESSEL			
	TOTAL SHEAR	PLATFORM SHEAR	VESSEL+ MISC. SHEAR	PIPE SHEAR	TOTAL SHEAR	PLATFORM SHEAR	VESSEL+ MISC. SHEAR	PIPE SHEAR
Average #1 - #13	54.6				67.1			
Based on ASCE 7-88 Simplified	55.6	(Note 4)	(Note 4)	(Note 4)	69.3	(Note 4)	(Note 4)	(Note 4)
Based on ASCE 7-88 Detailed	51.9	6.4	39.4	6.1	64.7	8.0	49.1	7.6
Based on ASCE 7-95 Simplified Category II	56.1	(Note 4)	(Note 4)	(Note 4)	68.0	(Note 4)	(Note 4)	(Note 4)
Based on ASCE 7-95 Detailed Category II	52.4	6.5	39.9	6.0	63.6	7.9	48.3	7.3
Based on ASCE 7-95 Simplified Category III	64.5	(Note 4)	(Note 4)	(Note 4)	78.2	(Note 4)	(Note 4)	(Note 4)
Based on ASCE 7-95 Detailed Category III	60.3	7.5	45.9	6.9	73.1	9.1	55.6	8.4

Notes to Table 3.3	
1	To convert kips to kilonewtons multiply values by 4.448.
2	No guidance provided for vertical vessels.
3	Load from platform at elevation 150 ft (45.72 m) only - Other platforms are included in shell loads.
4	Included in shell loads
5	Considers only forces generated above elevation 150 ft (45.72 m) - Remaining forces are included in shell loads

The criteria used for comparison of wind forces on vertical vessels are as follows:

- Circular platforms are 3 ft (0.91 m) wide measured from the edge of the tower.
- Vertical part of 18 in (0.46) diameter pipe acts as part of tower.
- Add 50% of shell weight to cover internals and miscellaneous.
- Assume vessel is 30% full of liquid (Specific Gravity = 1.0) during operation.

Table 3.3 presents the numerical comparison for each design practice. Base shears from wind forces varied from 48.7 to 65.3 kips (217 to 290 kN) for an empty vessel and 55.5 to 79.7 kips. (247 to 355 kN) for an operating vessel.

3.4.2 Horizontal Vessels

The same approach as vertical vessels was used to evaluate the loads on horizontal vessels. However, many of the design practices do not have explicit guidelines for horizontal vessels.

The horizontal vessel configuration used for the calculations is shown on Figure 3.6.

Table 3.4 presents the numerical comparison for each design practice. Base shears for transverse wind loading varied from 10.9 to 19.0 kips (48.5 to 84.5 kN).

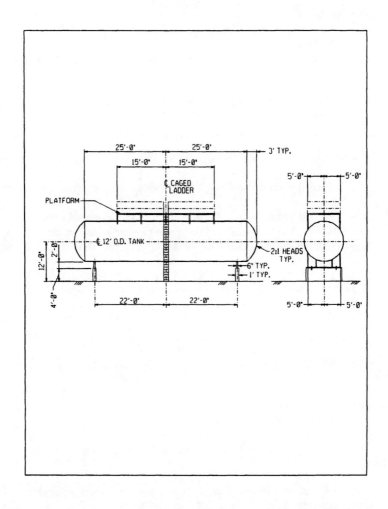

Figure 3.6 Example for Horizontal Vessel Comparison

Table 3.4 - Design Practice Comparison - Horizontal Vessel Wind Load

VALUES ARE BASE SHEAR IN KIPS
(Note 1)

DESIGN PRACTICE NO.	TRANSVERSE DIRECTION					LONGITUDINAL DIRECTION			
	Total Shear	Platform Shear	Ladder Shear	Support Shear	Vessel Shear	Total Shear	Platform Shear	Ladder Shear	Support Shear	Vessel Shear
#1	10.9	(Note 3)	(Note 3)	(Note 3)	10.9	2.2	(Note 3)	(Note 3)	(Note 3)	2.2
#2	19.0	4.3	(Note 3)	(Note 3)	14.7	6.6	4.3	(Note 3)	(Note 3)	2.3
#3	16.2	(Note 3)	(Note 3)	(Note 3)	16.2	4.4	(Note 3)	(Note 3)	1.6	2.8
#4	(Note 2)									
#5	(Note 2)									
#6	(Note 2)									
#7	(Note 2)									
#8	(Note 2)									
#9	19.0	5.9	1.3	0.2	11.6	(Note 4)				
#10	14.6	4.0	0.6	0.2	9.8	(Note 4)				
#11	14.6	4.0	0.6	0.2	9.8	(Note 4)				
#12	16.6	3.2	(Note 3)	0.8	12.6	(Note 4)				
#13	18.2	2.2	(Note 3)	(Note 3)	16.0	(Note 4)				

TABLE 3.4 - Design Practice Horizontal Vessel Wind Load Comparison (Cont'd)

VALUES ARE BASE SHEAR IN KIPS (Note 1)

AVERAGE & RECOMMENDED DESIGN PRACTICE	TRANSVERSE DIRECTION						LONGITUDINAL DIRECTION				
(SEE CHAPTER 4)	Total Shear	Platform Shear	Ladder Shear	Support Shear	Vessel Shear		Total Shear	Platform Shear	Ladder Shear	Support Shear	Vessel Shear
Average #1 - #13	16.1						4.4				
Based on *ASCE 7-88*	14.7	3.8	(Note 3)	0.3	10.6		8.3	1.3	(Note 3)	5.1	1.9
Based on *ASCE 7-95* Category II	15.83	4.09	(Note 3)	0.43	11.30		10.0	1.39	(Note 3)	6.61	2.0
Based on *ASCE 7-95* Category III	18.2	4.7	(Note 3)	0.5	13.0		11.5	1.6	(Note 3)	7.6	2.3

Notes to Table 3.4

1. To convert kips to kilonewtons multiply values by 4.448.
2. This practice does not address wind loads on horizontal vessels.
3. Included in shell loads based on increased diameter.
4. This practice does not address wind loads in the longitudinal direction.

CHAPTER 4
RECOMMENDED GUIDELINES

This study shows that the state-of-the-practice for the determination of wind loads on industrial structures is quite diverse. In many cases, the published theoretical and experimental work does not address petrochemical or industrial type structures. Therefore, the committee decided to provide recommended guidelines with a commentary. The committee has presented the guidelines in a manner such that the reader may reevaluate specific items in question.

Note: When used herein, *ASCE 7* refers to *ASCE 7-95*.

GUIDELINE	COMMENTARY
4.0 GENERAL	**C4.0 GENERAL**
Design wind forces for the main wind force resisting system and components should be determined by the equation $$F = q_z \, G \, C_f \, A \quad (4.1)$$ (where F is the applied wind force) using the following procedure:	The basic equation for design wind loading (Equation 4.1) is adopted from *ASCE 7* procedures for "Open Buildings and Other Structures" (*ASCE 7* Table 6-1). The provisions of Chapter 4 of this report primarily provide guidance in selecting appropriate force coefficients and projected areas.
1. A velocity pressure q_z is determined in accordance with the provisions of Section 6.5 of *ASCE 7*.	
2. The gust effect factor G (or G_f) is determined in accordance with the provisions of Table 6-1 and Section 6.6 of *ASCE 7*.	2. G_f is used in place of G for flexible structures, defined by *ASCE 7* as structures with a fundamental frequency f<1 Hz. If the height divided by least horizontal dimension is greater than 4,

a frequency check may be warranted.

ASCE 7 provides procedures for computing gust effect factors in the commentary section 6.6.

3. Force coefficients C_f and corresponding projected areas A_f or A_e are determined from the provisions of 4.1, 4.2, or 4.3 for pipe racks, open frame structures, and vessels respectively.

4.1 PIPE RACKS

Wind on the pipe rack structure itself should be calculated based on no shielding except as described in C4.1.1. For all structural members $C_f = 1.8$, or alternatively $C_f = 2.0$ below the first level and $C_f = 1.6$ for members above the first level.

4.1.1 Tributary Area for Piping

The tributary area for piping should be based on the diameter of the largest pipe plus 10% of the width of the pipe rack. This sum is multiplied by the length of the pipes (bent spacing) to determine the tributary area.

C4.1 PIPE RACKS

The C_f was determined with guidance from *ASCE 7*, Table 6-9 with consideration of typical solidities above and below the first level.

C4.1.1 Tributary Area for Piping

This area is based on the assumption that the wind will strike at an angle plus or minus from the horizontal with a slope of 1 to 10 and that the largest pipe is on the windward side. This corresponds to an angle of ±5.7 degrees. In some cases the pipe rack longitudinal strut or stringer might fall in the shielding envelope and should be deleted from wind load considerations.

This is a reasoned approach that accounts for wind on all the pipes (or cable trays) and shielding of the leeward pipes (or cable trays). The basis for the selection is a review of the existing practices. This effect is identified as needing further research by wind tunnel testing.

4.1.2 Tributary Area for Cable Trays

The tributary area for cable trays should be based on the height of the largest tray plus 10% of the width of the pipe rack. This sum is multiplied by the length of the trays (bent spacing) to determine the tributary area.

4.1.3 Force Coefficient for Pipes

The force coefficient $C_f = 0.7$ should be used as a minimum.

4.1.4 Force Coefficient for Cable Trays

For cable trays the force coefficient $C_f = 2.0$.

4.2 OPEN FRAME STRUCTURES

4.2.1 General

This section covers wind loads on open frame structures.

Wind loads should be calculated in accordance with the general procedures and provisions of *ASCE 7* for wind loads on "Other Structures" with the exceptions as noted.

C4.1.2 Tributary Area for Cable Trays

See commentary C4.1.1.

C4.1.3 Force Coefficient for Pipes

The force coefficient C_f, for pipe is taken from *ASCE 7*, Table 6-7 for a round shape, with h/D = 25, $D\sqrt{q_z} > 2.5$, and a moderately smooth surface; that is $C_f = 0.7$. If the largest pipe is insulated, then consider using a C_f for a rough pipe dependent on the roughness coefficient of the insulation (D'/D).

C4.1.4 Force Coefficient for Cable Trays

The force coefficient C_f, for cable trays is taken from *ASCE 7*, Table 6-7 for a square shape with the face normal to the wind and with h/D = 25; that is $C_f = 2.0$.

4.2.1.1 Main Wind Force Resisting System

1. Wind forces acting on the structural frame and appurtenances (ladders, handrails, stairs, etc.) should be computed in accordance with 4.2.2.

C4.2.1.1 Main Wind Force Resisting System

1. Alternatively, ladders, handrails and stairs can be treated as equipment instead of part of the main force resisting frame.

The basic method used to calculate wind loads on an open frame structure was adapted from a British method for computing wind forces on unclad framed buildings during construction (*Willford /Allsop*). That method covers simple three-dimensional rectangular frame structures with identical, regularly spaced frames in each direction made of sharp-edged members. It is based on theoretical work (*Cook*) and has been calibrated against the most extensive wind-tunnel test data available (*Georgiou 1979*). Thus, loads on the structure itself can be for a rectangular structure with similar frames using the methods of section 4.2.

The basic method has been extended to handle cases such as frames of unequal solidity, the presence of secondary beams (beams not along column lines), and frames made up of rounded members (*Willford/Allsop,Georgiou/Vickery/Church*). None of the extensions have been verified experimentally. However, it is still not unreasonable to presume that for a structure which is not particularly unusual, irregular, or having too many appurtenances, the procedures of 4.2

2. Wind forces on vessels, piping and cable trays located on or attached to the structure should be calculated according to the provisions of 4.1 and 4.3 and added to the wind forces acting on the frame in accordance with 4.2.6.

4.2.1.2 Force Coefficients for Components.

Wind loads for the design of individual components, cladding and appurtenances (excluding vessels, piping and cable trays) should be calculated according to the provisions of *ASCE 7*. Based on common practice force coefficients and areas for several items are given in Table 4.1.

4.2.2 Frame Load

Design wind forces for the main wind force resisting system for open frame structures should be determined by the equation:

$$F_s = q_z \, G \, C_f \, A_e \quad (4.1a)$$

should yield reasonably reliable wind loads for the structure and appurtenances together.

2. None of the theoretical or experimental work published to date has considered the inclusion of random three-dimensional solidity (e.g., vessels, heat exchangers, etc.) placed in the framework. However, it is expected that the total wind load on equipment will be less than the sum of the loads on the individual items due to shielding of, and by the frame, and also equipment to equipment shielding.

Thus, the approach taken in 4.2.6 is the reduction of the total wind load on equipment by a multiplication factor η to account for this shielding.

C4.2.2 Frame Load

The structure is idealized as two sets of orthogonal frames. The maximum wind force on each set of frames is calculated independently.

Note: C_f accounts for the entire structure in the direction of the wind.

TABLE 4.1 Force Coefficients for Wind Loads on Components

Item	C_f	Projected Area
Handrail	2.0	0.80 sq. ft./ft.
Ladder without cage	2.0	0.50 sq. ft./ft.
Ladder with cage	2.0	0.75 sq. ft./ft.
Solid Rectangles & Flat Plates	2.0	
Round or Square Shapes	See *ASCE 7* Table 6-7	
Stair w/handrail Side elevation	2.0	hand rail area plus channel depth
End elevation	2.0	50 % gross area

In Equation 4.1a F_s is the wind force on structural frame and appurtenances, q_z and G are as defined in 4.0, and

1. The force coefficient C_f is determined from the provisions of 4.2.3.

2. The area of application of force A_e is determined per 4.2.5.

3. The design load cases are computed per 4.2.6.

4.2.2.1 Limitations of Analytical Procedure.

Design wind forces are calculated for the structure as a whole.

The method is described for structures which are rectangular in plan and elevation.

C4.2.2.1 Limitations of Analytical Procedure

No information is provided about distribution of loads to individual frames. However, it should be noted that the windward frame will experience a much larger percentage of the total wind force than any other frames, except possibly for the case where the solidity ratio of the windward frame is much less than that of other frames.

4.2.3 Force Coefficients

The force coefficient for a set of frames shall be calculated by

$$C_f = C_{Dg} / \varepsilon \quad (4.2)$$

where

C_{Dg} is the force coefficient for the set of frames given in Figure 4.1, and
ε is the solidity ratio calculated in accordance with 4.2.4.

Alternately, C_f may be determined using Appendix 4.A.

Force coefficients are defined for wind forces acting normal to the frames irrespective of the actual wind direction.

C4.2.3 Force Coefficients

Force coefficients C_{Dg} are obtained from Figure 4.1 (see C4.2.1.1) or Appendix 4.A. A single value is obtained for each axis of the structure. This value is the maximum force coefficient for the component of force acting normal to the frames for all horizontal wind angles. Although the wind direction is nominally considered as being normal to the set of frames under consideration, the maximum force coefficient occurs when the wind is not normal to the frames (see C4.2.6.1 and 4A.1). The angle at which the maximum force coefficient occurs varies with the dimensions of the structure, the solidity, number of frames, and frame spacing.

A method to estimate this angle is given in the Appendix 4.A, which also provides C_f values for a larger range of S_F/B and ε values than Figure 4.1.

The force coefficients C_{Dg} were developed for use on the gross area (i.e., envelope area) of the structure as used by the British wind loading standard (*Willford/Allsop*). These are converted to force coefficients which are applied to solid areas as used in *ASCE 7* by Equation 4.2.

The force coefficients C_{Dg} were developed from wind tunnel tests for structures with a vertical aspect ratio (ratio of height to width perpendicular to the flow direction) of four. Although

Plan View Of Framing

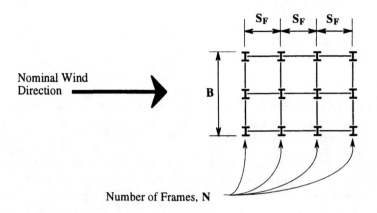

Notes:

(1) Frame spacing ratio is defined as S_F/B.
(2) Frame spacing, S_F, is measured from centerline to centerline.
(3) Frame width, B, is measured from outside edge to outside edge.
(4) Number of frames, N, is the number of framing lines normal to the nominal wind direction (N=4 as shown).
(5) Linear interpolation may be used for values of S_F/B not given on the following pages.

Figure 4.1 Force Coefficients, C_{Dg}, for Open Frame Structures

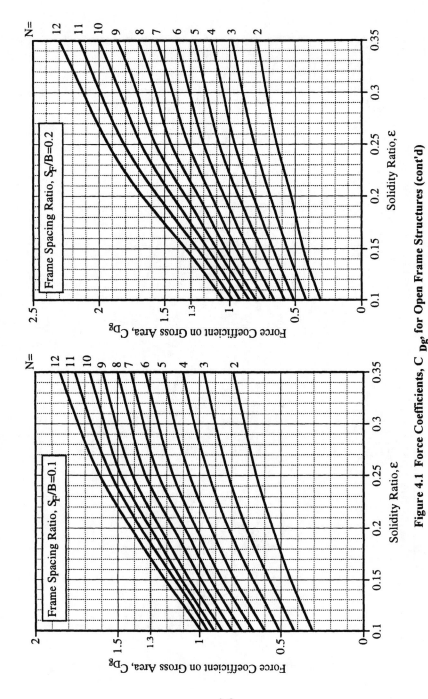

Figure 4.1 Force Coefficients, C_{Dg}, for Open Frame Structures (cont'd)

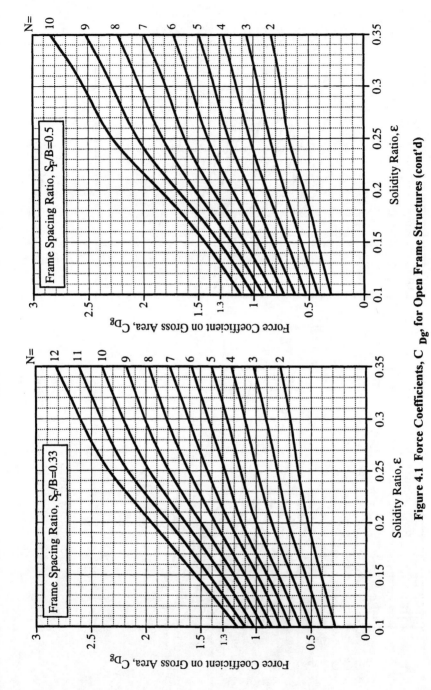

Figure 4.1 Force Coefficients, C_{Dg}, for Open Frame Structures (cont'd)

vertical aspect ratio does not play a large role in determining overall loads, the coefficients given in Figure 4.1 will likely be slightly conservative for relatively shorter structures and slightly unconservative for relatively taller structures

Force coefficients C_{Dg} are applicable for frames consisting of typical sharp-edged steel shapes such as wide flange shapes, channels and angles. Reference *Georgiou/Vickery/Church* suggests a method to account for structures containing some members of circular or other cross sectional shape.

4.2.4 Solidity Ratio

The solidity ratio ε is given by;

$$\varepsilon = A_s / A_g \qquad (4.3)$$

where A_g is the gross area (envelope area) of the windward frame and A_s is the effective solid area of the windward frame defined by the following :

4.2.4.1 The solid area of a frame is defined as the solid area of each element in the plane of the frame projected normal to the nominal wind direction. Elements considered as part of the solid area of a frame include beams, columns, bracing, cladding, stairs, ladders, handrails, etc. Items such as vessels, tanks, piping and cable trays are not included in calculation of solid area of frame; wind loads on these items are calculated separately.

4.2.4.2 The presence of flooring or decking does not cause an increase of the

C4.2.4 Solidity Ratio

Reference *Willford/Allsop* presents a method to account for the effects of secondary floor beams (beams not in the plane of a frame). Use of this method may result in a small increase in the total wind force on the structure. With the associated uncertainties with the determination of the wind forces this minor addition may be ignored.

C4.2.4.2 Reference *Willford/Allsop* indicates that although extremely little

solid area of 4.2.4.1 beyond the inclusion of the thickness of the deck.

experimental work has been done regarding effects of flooring, the limited data available suggest that the presence of solid decking does not increase wind forces above those calculated for the bare frame, and may in fact reduce the loads due to a "streamlining" effect. No research relating to open grating floors has been published. The opinion of the committee is that open grating floors will not significantly affect the wind forces on the structure

4.2.4.3 For structures with frames of equal solidity, the effective solid area A_S should be taken as solid area of the windward frame.

4.2.4.4 For structures where the solid area of the windward frame exceeds the solid area of the other frames, the effective solid area A_S should be taken as the solid area of the windward frame.

C4.2.4.4 The force coefficients of Figure 4.1 were developed for sets of identical frames. Research shows that the solidity of the windward frame is the most critical (*Cook, Whitbread*), leading to the recommendation. This provision is likely to yield slightly conservative loads, since the greater the solidity of the windward frame with respect to the other frames, the greater the shielding of the other frames.

4.2.4.5 For structures where the solid area of the windward frame is less than the solid area of the other frames, the effective solid area A_S should be taken as the average of all the frames.

4.2.5 Area of Application of Force

A_e shall be calculated in the same manner as the effective solid area in 4.2.4 except that it is for the portion of the structure height consistent with the velocity pressure q_z.

4.2.6 Design Load Cases

The total wind force acting on the structure in a given direction, F_T, is equal to the sum of the wind loads acting on the structure and appurtenances (F_S), plus the wind load on the equipment and vessels (per 4.3), plus the wind load on piping. See Figure 4.2 for complete definitions of F_T and F_S

If piping arrangements are not known the engineer may assume the piping area to be 10% of the gross area of the face of the structure for each principal axis. A force coefficient of 0.7 should be used for this piping area.

The following two load cases must be considered as a minimum.

4.2.6.1 Frame load + equipment load + piping load (F_T) for one axis, acting simultaneously with 50% of the frame load (F_S) along other axis, for each direction. These two combinations are indicated in Figure 4.2.

C4.2.6 Design Load Cases

In some cases this design load will exceed the load which would occur if the structure were fully clad. It is also possible that the wind load on just the frame itself (before equipment loads are added) will exceed the load on the fully clad structure. This happens most often for structures with at least 4 to 5 frames and relatively higher solidities. This phenomenon is very clearly demonstrated in *Walshe*, which presents force coefficients on a building for 10 different stages of erection, from open frames to the partially clad to then fully clad building. The wind load on the model when fully clad is less than that during several stages of erection.

C4.2.6.1 While the maximum wind load normal to the frame for a structure consisting of a single frame occurs when the wind direction is normal to the plane of the frame, this is not the case for a multiple open framed structure. Maximum load normal to the plane of the frames occurs when the wind direction is typically 10 to 45 degrees from the normal (*Willford/Allsop*). This is due to the fact that for oblique winds there is no direct shielding of successive columns and because a larger area of frame is exposed to the wind directly (without shielding) as the wind angle increases. Thus the maximum wind load on one set

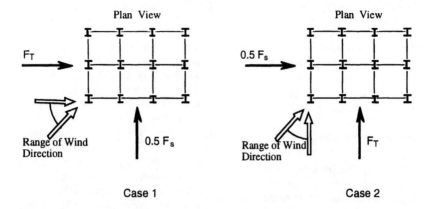

Notes:

(1) F_s denotes the wind force on the structural frame and appurtenances in the indicated direction (excludes wind load on equipment, piping and cable trays).

(2) F_T denotes the total wind force on the structure in the direction indicated, which is the sum of the forces on the structural frame and appurtenances, equipment, and piping. If appropriate, the equipment load may be reduced by considering shielding effects per 4.2.6.2.

(3) Load combination factors applied to F_s may alternately be determined by the detailed method of Appendix 4A and used in place of 0.5 values shown. These values shall be calculated separately for Case 1 and Case 2.

Figure 4.2 Design Load Cases

of frames occurs at an angle which will also induce significant loads on the other set of frames (*Willford/Allsop, Georgiou/Vickery/Church*).

Full and partial loading of structures given in *ASCE 7* section 6.8 were developed for clad structures only. The provisions of that paragraph are not applicable to open frame structures due to the different flow characteristics

4.2.6.2 When, in the engineer's judgment, there is substantial shielding of equipment by the structure or other equipment on a given level in the wind direction under consideration, the wind load on equipment in 4.2.6.1 on that level may be reduced by the shielding factor η.

$$\eta = (1 - \varepsilon)^{(\kappa + 0.3)} \qquad (4\text{-}4)$$

where $\eta \geq 0.4$.

The solidity ratio ε is defined in 4.2.4.

κ is the volumetric solidity ratio for the level under consideration, defined as the ratio of the sum of the volumes of all equipment, vessels, exchangers, etc. on a level of the structure to the gross volume of the structure at that level.

κ should be taken as 0 when there is no equipment to equipment shielding (e.g., if there is only one exchanger or vessel on the level under consideration, or the equipment is widely spaced).

The wind load on any equipment or portion thereof which extends above the top of the structure should not be reduced.

C4.2.6.2 These provisions are an attempt by the committee to recognize the beneficial effects of shielding of equipment by the structure and other equipment. The form is very loosely based on shielding equations developed for sets of trusses, with an additional factor introduced to account for the presence of solid elements.

The factor κ is used to account for equipment to equipment, and equipment to structure shielding.

4.2.6.3 Horizontal Torsion

Horizontal torsion (torsion about the vertical axis) may be a factor for open frame structures. The engineer should consider the possibility of torsion in the design

Consideration should be given to the application points of the wind load, especially in cases where the building framing is irregular and/or equipment locations are not symmetric.

4.3 PRESSURE VESSELS

Where vessel and piping diameters are specified, it is intended that insulation, if present, be included in the projected area. Insulation should not be included for stiffness when checking h/D for dynamic characteristics.

C4.3 PRESSURE VESSELS

For tall slender vessels, vortex shedding may cause significant oscillating force in the crosswind direction. This means that the structure may experience significant loads in both the alongwind and crosswind directions at the same time. Crosswind forces such as vortex shedding are not addressed in this document.

4.3.1 Vertical Vessels

4.3.1.1 Use *ASCE 7* to calculate velocity pressures and to obtain gust effect factors.

4.3.1.2 Simplified Method

C.4.3.1.2 Simplified Method

If detailed information (number of platforms, platform size, etc.) is unknown at the time of design of the foundation/piles, the following approach may be used:

1) For the projected width, add 5-ft. (1.52 m) to the diameter of the vessel, or add 3-ft. (0.91 m) plus the diameter of the largest pipe to the diameter of the vessel, whichever is greater. This will account for platforms, ladders, nozzles and piping below the top tangent line.

2) The vessel height should be increased one (1) vessel diameter to account for a large diameter pipe and platform attached above the top tangent, as is the case with most tower arrangements.

3) The increases in vessel height or diameter to account for wind on appurtenances should not be used in calculating the h/D ratio for force coefficients or flexibility.

4) The force coefficient (C_f) should be determined from *ASCE 7*, Table 6-7.

1) With limited information on the vessel and appurtenances, this simple approach gives reasonably consistent results.

2) This is an approximation to alleviate the need for some rather tedious calculations based on gross assumptions.

3) As noted in 4.3 previously, insulation should not be included in the h/D calculations.

4) Any roughness due to nozzles, ladders and other appurtenances is covered by the increase in diameter.

A moderately smooth surface should normally be assumed. If ribbed insulation will be used then the D'/D should be calculated.

4.3.1.3 Detailed Method

If most design detail items (platforms, piping, ladders, etc.) of the vessel are known, the following method should be used:

1) For the projected width, add 1.5-ft. (0.46 m) to the vessel diameter to account for ladders, nozzles and piping 8-in. (0.2 m) or smaller and add the diameter of the largest line coming from the top portion of the vessel.

2) The force coefficient (C_f) should be

C4.3.1.3 Detailed Method

This will provide more accurate values for foundation design.

1) This is consistent with the values most companies are using.

taken from *ASCE 7*, Table 6-7 based on appropriate roughness at vessel surface.

3) For pipes outside the projected width of the vessel (defined in 1) larger than 8-in. (0.2 m), including insulation, use the projected area of the pipe and use a force coefficient (C_f) of 0.7.

For pipes inside the projected width of the vessel (defined in 1) larger than 8-in. (0.2 m), including insulation, and more than 5 pipe diameters from the vessel surface, add the projected area of the pipe and use a force coefficient (C_f) of 0.7.

3) C_f is determined from *ASCE 7*, Table 6-7 for a moderately smooth surface.

4) For platforms, use the projected area of the support steel and a force coefficient (C_f) of 2.0.

For handrails use the values for area and force coefficient from Table 4.1.

Where the railing projects beyond the vessel, the projected area of two (2) sets of railing systems should be used. A front system and a back system should both be projected.

4) The front and back systems of railings are far enough apart to preclude shielding.

4.3.2 Horizontal Vessels

4.3.2.1 No check for dynamic properties is required.

4.3.2.2 For the projected diameter, add 1.5 ft. (0.46 m) to the insulated diameter to account for ladders, nozzles and pipe 8 in. (0.2 m) (including insulation) or smaller.

4.3.2.3 For wind perpendicular to the long axis of the vessel (transverse wind), the force coefficient (C_f) should be determined from *ASCE 7*, Table 6-7.

C4.3.2.3 Use B/D to determine C_f similar to h/D for vertical vessels.

4-18

4.3.2.4 For wind in the longitudinal direction, use C_f of 0.5 for a rounded head and 1.2 for a flat head.

C4.3.2.4 This wind direction will seldom control design of foundations.

4.3.2.5 For pipe larger than 8 in. (0.2 m), including insulation, use the projected area of the pipe and use a force coefficient (C_f) of 0.7.

4.3.2.6 For platforms, use the projected area of the support steel and a force coefficient (C_f) of 2.0.

For handrails use the values for area and force coefficient from Table 4.1.

Use the projected area of each railing system.

C4.3.2.6 The reason for projecting the front and back railing system is that they are far enough apart to preclude shielding.

4.3.2.7 For supports use the actual projected area. C_f should be 1.3 for concrete pedestals. For steel supports, use the method described for platforms.

C4.3.2.7 The 1.3 factor is used because a pedestal is similar to a bluff rectangular body.

4.3.3 Spheres

4.3.3.1 No check for dynamic properties is required.

4.3.3.2 For the projected diameter, add 1.5 ft. (0.46 m) to the insulated diameter to account for ladders, nozzles and pipe 8 in. (0.2 m) (including insulation) or smaller.

4.3.3.3 Use $C_f = 0.5$ (for vessel only). Supports should be evaluated separately.

C4.3.3.3 See *ASCE Wind*.

4.3.3.4 For pipe larger than 8 in. (0.2 m), including insulation, use the projected area of the pipe and use a force coefficient (C_f) of 0.7.

C4.3.3.4 C_f is determined from *ASCE 7*, Table 6-7 for a moderately smooth surface.

4.3.3.5 For platforms, use the projected area for the support steel and a force coefficient (C_f) of 2.0.

For handrails use the values for area and force coefficient from Table 4.1.

Use the projected area of each railing system.

4.3.3.6 For supports use the actual projected area. C_f should be 1.3 for rectangular concrete columns and 0.7 for circular columns. For steel supports, use the method described for platforms.

C4.3.3.5 The reason for projecting the front and back railing system is that they are far enough apart to preclude shielding.

APPENDIX 4.A
ALTERNATE METHOD FOR DETERMINING C_F AND LOAD COMBINATIONS FOR OPEN FRAME STRUCTURES

4A.1 BACKGROUND

Maximum wind force normal to the face of a rectangular enclosed building occurs when the wind direction is normal to the building face. The same is true for wind load on a single frame or solid sign. However, this is not the case for an open frame structure with more than one frame. As the wind direction moves away from the normal and more towards a quartering wind, columns which once lined up neatly behind each other, shielding each other, become staggered and exposed to the full wind. Additionally, the area of the structure projected on a plane normal to the wind also increases.

The variation of the wind loads along each principal axis of a rectangular open frame structure with the direction of wind is shown in Figure 4A.1, for the structure and wind angle of attack defined in Figure 4A.2. It can readily be seen that when one frame set experiences its maximum frame load 'A' or 'D', the frame set along the other axis experiences a wind force 'C' or 'B' respectively, thus the need for the load combinations of Section 4.2.6.1. In those provisions, the load at 'C' is roughly estimated to be 50% of 'A' and 'B' is estimated to be 50% of the load at 'D'. In actuality, the loads on the secondary axis can range from about 25% to 75% of the primary axis load, depending on many factors including spacing ratio, number of frames, solidity ratio, etc. This appendix provides a method to obtain a better estimate of the simultaneously acting load on the secondary axis.

4A.2 FORCE COEFFICIENTS

This method provides force coefficients C_f for a greater range of ε and S_F/B values than the method of 4.2.3, as well as providing an estimate of α_{max}. For cases where both methods are applicable, they will generally yield very similar results. References *Nadeem* and *Nadeem/Levitan* discuss this method in greater detail. The procedure is as follows:

1. Determine ε, S_F/B, and N for the principal axis under consideration as per 4.2.4 and Figure 4.1.

2. Estimate the wind angle of attack which maximizes the force parallel to the axis under consideration.

$$\alpha_{max} = (10 + 58\varepsilon)° \quad \text{for } 3 \le N \le 5 \quad (4A.1)$$
$$\alpha_{max} = (16 + 52\varepsilon)° \quad \text{for } 6 \le N \le 10 \quad (4A.2)$$

3. Estimate the force coefficient C_f from Figure 4A.3 by the following procedure:

   a) Determine C_f from Figure 4A.3(a) for wind angle of attack = α_{max} and appropriate spacing ratio S_F/B,. This C_f is for a structure with N=3 frames and a solidity ratio of ε =0.1.

   b) Determine C_f from Figure 4A.3(b) for a structure with N=3 frames and a solidity ratio of ε =0.5.

   c) Interpolate between results of a) and b) for the actual solidity ratio, yielding a force coefficient for the correct spacing and solidity ratios, and N=3 frames, $C_{f,N=3}$.

   d) Determine C_f from Figure 4A.3(c) for a structure with N=10 frames and a solidity ratio of ε =0.1.

   e) Determine C_f from Figure 4A.3(d) for a structure with N=10 frames and a solidity ratio of ε =0.5.

   f) Interpolate between results of d) and e) for the actual solidity ratio, yielding a force coefficient for the correct spacing and solidity ratios, and N=10 frames, $C_{f,N=10}$.

   g) Determine C_f for the axis under consideration by interpolating between $C_{f,N=3}$ and $C_{f,N=10}$ for the actual number of frames

   Note that if the structure has exactly 3 or 10 frames, only steps (a-c) or (d-f) respectively need be used. Similarly, if a structure has a solidity ratio very near to 0.1 or 0.5, only one interpolation between Figures 4A.3(a) and (c) or 4A.3(b) and (d) respectively would be necessary.

4A.3 LOAD COMBINATIONS

Section 4.2.6.1 specified the load combination of full wind load on the axis under consideration acting simultaneously with 50% of the frame wind load on the other axis. A more detailed method to estimate the wind load acting simultaneously on the secondary axis frames is given here.

1. Determine C_f for the principal axis under consideration as per 4.2.3 or 4A.2. If provisions of 4.2.3 are used, α_{max} must still be determined as per 4A.2

2. Determine the force coefficient C_f for the secondary axis from Figure 4A.3, using ε, S_F/B, and N values for the secondary axis and a wind angle of attack of (90°- α_{max}). Step 3 in section 4A.2 explains how to obtain C_f from Figure 4A.3.

4A.4 SAMPLE CALCULATIONS

The proposed method has been used to calculate the force coefficients for a structure whose plan is shown in Figure 4A.4. Summarizing the important frame set properties,

For winds nominally from west to east
(i.e., winds normal to the N-S frame set)
$\varepsilon = 0.136$
$N = 4$
$S_F = 25.4$ ft (7.75 m)
$B = 68.9$ ft (21.0 m)
$S_F/B = 25.4 / 68.9 = 0.369$

For winds nominally from south to north
(i.e., winds normal to the E-W frame set)
$\varepsilon = 0.286$
$N = 5$
$S_F = 16.4$ ft (5.0 m)
$B = 78.75$ ft (24.0 m)
$S_F/B = 16.4 / 78.75 = 0.208$.

Determining Force Coefficients: To determine the force coefficient for the E-W structural axis (winds nominally normal to the N-S frame), first estimate the wind angle of attack at which this maximum load will occur. Since N=4 and $\varepsilon = 0.136$ for the N-S frame set, Equation 4A.1 yields

$\alpha_{max} = 10 + 58(0.136) = 18°$.

From Figures 4A.3(a) and (b), for $S_F/B = 0.369$, $C_f = 3.87$ and 2.10 for structures with $\varepsilon = 0.1$ and $\varepsilon = 0.5$ respectively and N=3 frames. Interpolating between these two values for $\varepsilon = 0.136$,

$C_{f,N=3} = 3.87 - [(3.87 - 2.10)/0.4](0.136 - 0.1) = 3.71$.

From Figures 4A.3(c) and (d), for $S_F/B = 0.208$, $C_f = 10.08$ and 3.15 for structures with $\varepsilon = 0.1$ and $\varepsilon = 0.5$ respectively and 10 frames. Interpolating between these two values for $\varepsilon = 0.136$,

$$C_{f,N=10} = 10.08 - [(10.08 - 3.15)/0.4](0.136 - 0.1) = 9.46.$$

Interpolating between the two previous results of $C_{f,N=3} = 3.71$ and $C_{f,N=10} = 9.46$ for the case $N = 4$,

$$C_f = 3.71 + [(9.46 - 3.71)/7](4 - 3) = 4.53$$

gives the maximum force coefficient for the N-S set of frames, occurring near $\alpha_{max} = 18°$.

Determining Load Combinations: While the maximum wind load is acting on the N-S set of frames, the wind simultaneously acts on the E-W set of frames at angle of attack of $90° - 18° = 72°$.

The force coefficient for the E-W frames is determined as per step 2 of 4A.3.
For $\alpha = 72°$, $S_F/B=0.208$ and $\varepsilon = 0.286$, interpolation between Figures 4A.3(a) and (b) yields $C_{f,N=3} = 0.91$. Interpolation between Figures 4A.3(c) and (d) yields $C_{f,N=10} = 2.84$.

One more interpolation between $C_{f,N=3}$ and $C_{f,N=10}$ for $N= 5$ frames yields $C_f = 1.46$. which is the force coefficient for the wind load acting on the E-W frame set while the N-S set is experiencing it's maximum wind load. This combination of loads is shown in Figure 4A.5(a).

This entire procedure should now be repeated assuming that maximum wind load acts on the E-W set of frames. For this case, $\alpha_{max} = 27°$, $C_f = 4.0$ for the E-W frames, with a simultaneously acting load of $C_f = 2.19$ for the N-S frame set, as shown in Figure 4A.5(b).

Note that for the case of full wind load on the E-W axis, use of this alternate procedure reduced the wind load acting simultaneously on the N-S axis from 50% (4.2.6.2) to $1.46/4.00 = 37\%$. The load combination for full wind on the N-S structure axis remained was close to the recommended 50% at $2.19/4.53 = 48\%$.

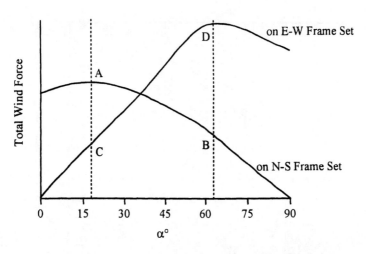

Figure 4A.1 Variation of Wind Load vs. Wind Direction

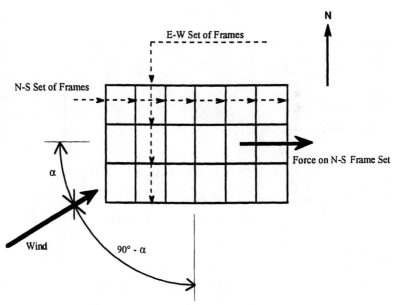

Figure 4A.2 Plan View of Structural Framing

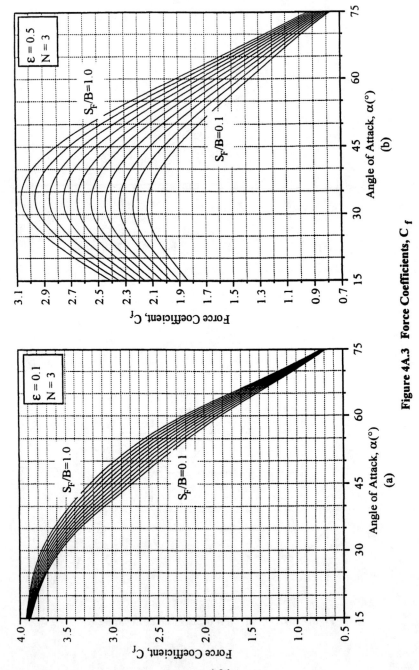

Figure 4A.3 Force Coefficients, C_f

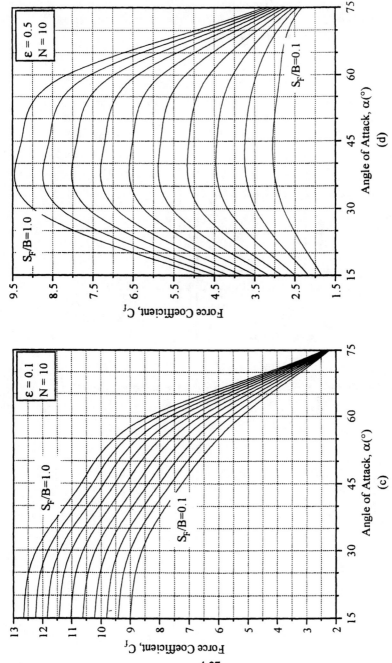

Figure 4A.3 Force Coefficients, C_f (cont'd)

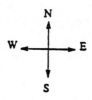

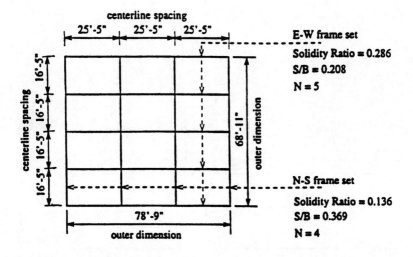

Figure 4A.4 Structural Framing Plan for Sample Calculations

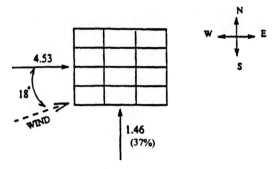

(a) Considering Maximum Load on N-S Frame Set

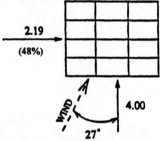

(b) Considering Maximum Load on E-W Frame Set

Figure 4A.5 Force Coefficients, C_f, for Example Design Load Combinations

CHAPTER 5
EXAMPLES

This chapter has three appendixes 5.A, 5.B and 5.C will demonstrate the application of the recommended guidelines for calculating wind loads on pipe racks, open frame structures, and vessels. The structures considered will be based on the ones described in Chapter 3. The structures are located in an assumed petrochemical plant near Lake Charles, Louisiana (in southwest Louisiana about 20 miles from the coast).

All calculations are performed using the guidelines from Chapter 4 and the provisions of *ASCE* 7-95. Design wind forces for all of the structures are determined using the expression

$$F = q_z \, G \, C_f \, A \qquad (4.1)$$

The velocity pressure q_z is determined using Section 6.5 of *ASCE* 7.

$$q_z = 0.00256 \, K_z \, K_{zt} \, V^2 \, I \qquad (ASCE\ 7,\text{Eq. 6-1})$$

The basic wind speed V is obtained from Figure 6-1 of *ASCE* 7. For the plant location near Lake Charles, Louisiana, V≈120 mph (54 m/s). This is a 3-second gust wind speed with an annual probability of exceedence of 0.02 (i.e., mean recurrence interval of 50 years).

The importance factor I is obtained from *ASCE* 7 Table 6-2. For this example an importance factor was chosen based on Category III structures.

The exposure category is selected in accordance with *ASCE* 7, Section 6.5.3. Exposure C, open terrain with scattered obstructions, was used for this example. Assuming no terrain features such as hills or escarpments are present, the topographic factor K_{zt}=1.0 per *ASCE* 7 section 6.5.5. The velocity pressure exposure coefficients K_z are obtained from *ASCE* 7, Table 6-3. Velocity pressures are determined using *ASCE* 7, Eq. 6-1 shown above. Table 5.1 gives values for q_z at several heights.

The gust effect factor G is determined in accordance with the provisions of Table 6-1 and Section 6.6 of *ASCE 7*. For rigid structures, the simplified method of section 6.6.1 specifies G=0.85 for structures in terrain exposure C. A more detailed analysis method for rigid structures is presented in the *ASCE 7* Commentary section 6.6. This detailed method is appropriate for very large structures and where more accuracy is desired. For flexible or dynamically sensitive structures, defined as those structures with a fundamental frequency $f < 1$ Hz (i.e., fundamental period of vibration > 1 second), G_f is used in place of G. A method to determine G_f is given in *ASCE 7* Commentary section 6.6.

TABLE 5.1
Velocity Pressure Profile for Examples

Height above Ground z (ft)	K_z	Velocity Pressure q_z (psf)
0-15	0.85	36.0
20	0.90	38.2
25	0.94	39.8
30	0.98	41.5
40	1.04	44.1
50	1.09	46.2
60	1.13	47.9
70	1.17	49.6
80	1.21	51.3
90	1.24	52.6
100	1.26	53.4
120	1.31	55.5
140	1.36	57.7
160	1.39	58.9

Note : To convert psf to N/m2 multiply values in the table by 47.878.

APPENDIX 5.A
PIPE RACK EXAMPLE

The pipe rack cases considered will be as described in Section 3.2, Figure 3.1 and further below. Design wind forces are determined by Equation 4.1a (repeated below) where F is the force per unit length of the piping or cable tray:

$$F = q_z G C_f A_e$$

Design wind pressure, for 30 ft elevation from Table 5.1

$$q_z = 41.5 \text{ psf } (1.99 \text{ kN/m}^2)$$

Gust effect factor, $G = 0.85$ (*ASCE 7*, Section 6.6.1)

Force Coefficients

    For structural members, $C_f = 1.8$ or alternatively for structural members above the first level $C_f = 1.6$ and below the first level $C_f = 2.0$.   (Section 4.1)
    For pipes, $C_f = 0.7$ (Section 4.1.3)
    For cable trays, $C_f = 2.0$ (Section 4.1.4)

Projected Area

Projected Area per foot of pipe rack, A_e = Largest pipe diameter or cable tray height + 10% of the rack width.
                                                                             (Sections 4.1.1 and 4.1.2)

5A.1 PART I - PIPING AND CABLE TRAY

The guidelines require the consideration of the piping or cable tray separately from the structural members. The following calculations are only for piping or cable trays without the structural support members.

Case I - 20 ft (6.10 m) Wide Rack with one (1) - 48 in (1.22 m) Pipe and fourteen (14) -9 in (0.23 m) Pipes

Projected Area, A_e = 4 ft + (20 ft x 0.10) = 6.0 ft^2/ft (1.83 m^2/m)
Force per foot F = 41.5 psf x 0.85 x 0.7 x 6.0 ft^2/ft = 148 plf (2.16 kN/m)

Case II - 20 ft (6.10 m) Wide Rack with fifteen (15) - 12 in (0.30 m) Pipes

Projected Area, A_e = 1 ft + (20 ft x 0.10) = 3 ft^2/ft (0.92 m^2/m)
Force per foot F = 41.5 psf x 0.85 x 0.7 x 3.0 ft^2/ft = 74 plf (1.08 kN/m)

Case III - 5 ft 6 in (1.68 m) Wide Rack with one (1) - 24 in (0.61 m) pipe and two (2) - 12 in (0.30 m) Pipes

Projected Area, A_e = 2 ft + (5.5 ft x 0.10) = 2.55 ft^2/ft (0.77 m^2/m)
Force per foot F = 41.5 psf x 0.85 x 0.7 x 2.55 ft^2/ft = 63 plf (0.92 kN/m)

Case IV - 4 ft 6 in (1.37 m) Wide Rack with three (3) - 12 in (0.30 m) Pipes

Projected Area, A_e = 1 ft + (4.5 ft x 0.10) = 1.45 ft^2/ft (0.44 m^2/m)
Force per foot F = 41.5 psf x 0.85 x 0.7 x 1.45 ft^2/ft = 36 plf (0.52 kN/m)

Case V - 20 ft (6.10 m) Wide Rack with one (1) - 36 in (0.91 m) Pipe, two (2) - 24 in (0.61 m) Pipes, four (4) - 12 in (0.30 m) Pipes, and six (6) - 9 in (0.23 m) Pipes

Projected Area, A_e = 3 ft + (20 ft x 0.10) = 5.0 ft^2/ft (1.53 m^2/m)
Force per foot F = 41.5 psf x 0.85 x 0.7 x 5.0 ft^2/ft = 123 plf (1.79 kN/m)

Cable Trays- 20 ft (6.10 m) Wide Rack with two (2) - 36 in (0.91 m) Trays, one (1) - 24 in (0.61 m) Tray, two (2)- 18 in (0.46 m) Trays, two (2) - 12 in (0.30 m) Trays, and two (2) - 6 in (0.15 m) all 6 in (0.15 m) high

Projected Area, A_e = 0.5 ft + (20 ft x 0.10) = 2.5 ft^2/ft (0.76 m^2/m)
Force per foot F = 41.5 psf x 0.85 x 2.0 x 2.5 ft^2/ft = 176 plf (2.57 kN/m)

5A.2 PART II - STRUCTURAL MEMBERS

For structural members assume the pipe rack geometry is as follows (see Figure 5A.1):

- 20 ft (6.10 m) wide rack with bent spacing on 20 ft (6.10 m) centers all stringers not shielded.

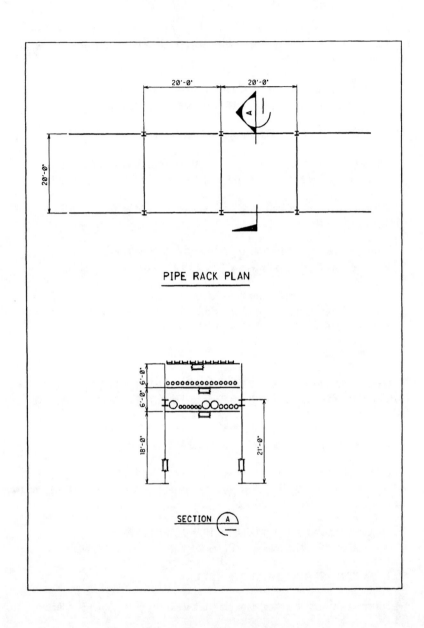

FIGURE 5A.1 Pipe Rack Example

- The stringer and the columns are assumed to be W10 and W12 sections respectively. Both stringer and column are assumed to have fireproofing insulation.
- Stringer member projected area per foot of rack = 1 ft²/ft (0.31 m²/m)
- Column projected area per foot of column = 1.25 ft²/ft (0.38 m²/m)
- Three levels of pipes and cable trays at elevations 18 ft (5.49 m), 24 ft (7.32 m), and 30 ft (9.15 m)
- One level of struts at 21 ft (6.40 m)
- Conservatively, q_z [at elevation 30 ft (9.15 m)] = 41.5 psf (1.91 kN/m²) for all members.

Next calculate wind loads per bent using a force coefficient C_f = 1.8 for all members per Section 4.1. Projected area for stringers at elevation 21 ft (6.40 m) is calculated as the sum of stringers times the stringer depth times the bent spacing.

Projected Area of Stringers = 2 stringers x 1 ft depth x 20 ft bent spacing
= 40 ft² (3.72 m²)

Similarly the projected area for columns is:

Projected Area of Columns = 2 columns x 1.25 ft width x 30 ft height = 75 ft²
(6.97 m²)

Total force on the structural members is per Equation 4.1:

$F = q_z\ G\ C_f\ A_e$ = 41.5 psf x 0.85 x 1.8 x (40 ft² + 75 ft²) = 7300 lbs (32.5 kN)

Alternatively, using C_f = 1.6 for members above the first level and C_f = 2.0 for members a below the first level.

Projected area for stringers = 40 ft² (3.72 m²)

Projected area for columns above first level = 2 columns x 1.25 ft x 12 ft high
= 30 ft² (2.79 m²)

Projected area for columns below first level = 2 columns x 1.25 ft x 18 ft high
= 45 ft² (4.18 m²)
Total force on structure F = 41.5 psf x 0.85 x [(40 ft² + 30 ft²) x 1.6 + 45 ft² x 2.0]
= 7125 lbs (31.7 kN)

5A.3 PIPE RACK EXAMPLE - SUMMARY AND CONCLUSION

To combine the effects of the piping, cable tray and the structural members, the pipe rack structure along with Case V pipes on the bottom level, Case II pipes on the middle level, and cable trays on the top level are used. See Figure 5A.1.

Continued on next page

Calculate the Total Base Shear using $C_f = 1.8$ for the structure.

Bottom Level of Pipe	123 plf x 20 ft	=	2460 lbs (11.0 kN)
Middle Level of Pipe	74 plf x 20 ft	=	1480 lbs (6.6 kN)
Top Level of Cable Trays	176 plf x 20 ft	=	3520 lbs (15.7 kN)
Structure		=	7300 lbs (32.5 kN)
Total Base Shear per bent		=	14760 lbs (65.8 kN)

APPENDIX 5.B
OPEN FRAME EXAMPLE

The structure considered will be the one described in Section 3.3 and shown in Figures 3.2, 3.3 and 3.4. Design wind forces are determined by Equation 4.1a;

$F = q_z GC_f A_e$

It is convenient to determine the velocity pressures at the mid-floor heights and at the top of the structure. From Table 5.1, q_z and K_z are determined and summarized in Table 5B.1.

TABLE 5B.1
q_z and K_z

Height above Ground z (ft)	K_z	q_z (psf)
10	0.85	36.0
34	1.00	42.4
65	1.15	48.8
h = 83	1.22	51.7

Note : To convert psf to N/m^2 multiply values in this table by 47.878.

Although the top of the third floor level is at 82 ft, the structure height h was increased slightly to account for the handrail and minor equipment on top of the structure (see Figures. 3.3 and 3.4).

The gust effect factor is determined next. The ratio of height/least horizontal dimension = 83 ft / 41 ft = 2.02 < 4, therefore the structure is not considered a flexible structure. Use G = 0.85, as described at the beginning of Chapter 5.

5B.1 ALONG WIND FORCE CALCULATIONS

In order to calculate the force coefficient, the solidity ratio ε must first be computed from Equation 4.3, which stated $\varepsilon = A_s / A_g$. The gross area (or envelope area) is the area within the outmost projections of the front face normal to the nominal wind direction. Note that the width used below is measured from outside column face to outside column face. For the wind direction shown in Figure 3.2,

$$A_g = 83 \text{ ft (height)} \times 41 \text{ ft (width)} = 3{,}403 \text{ ft}^2 \quad (316 \text{ m}^2)$$

To determine the effective solid area, the solid area of the windward frame must first be calculated per 4.2.4.1. In order to facilitate the computation of forces later in the problem, it is convenient to calculate the solid areas from mid-floor to mid-floor, and sum these to obtain the total solid area of the frame. Member sizes are given in Section 3.3 (except details of handrails and stairs). Calculation of the solid area of the windward frame (column line 3) is summarized in Table 5B.2. The stairs were considered as part of the windward frame (see Figure 3.2). The stairs column in Table 5B.2 includes areas of stair stringers, struts, handrails, and bracing.

TABLE 5B.2
Solid Area of Windward Frame - A_s

Floor Level	Tributary Height (ft)	Solid Areas (ft²)						
		Cols.	Beams	Int. Beams	Bracing	Hand Rails	Stairs	Total
0	0-10	30	0	0	19	0	76	125
1	10-34	72	120	40	40	32	150	454
2	34-65	93	60	80	38	40	91	402
3	65-83	51	60	40	28	40	17	236
Total Solid Area of Windward Frame (ft²) =								1217 (113 m²)

Note : To convert ft² to m² multiply values in this table by 0.0929.

Since the middle and leeward frames (column lines 2 and 1, respectively) are similar to the windward frame with the exception of not having stairs, the solid areas and hence the solidity ratios for these two frames will be less than the windward frame, so A_s is equal to the solid area of windward frame per 4.2.4.4, which leads to

$$\varepsilon = A_s / A_g = 1{,}217 \text{ ft}^2 / 3{,}403 \text{ ft}^2 = 0.358$$

Next, the coefficient C_{Dg} is obtained from curves given in Figure 4.1 as a function of the solidity ratio ε, the number of frames N, and the frame spacing ratio

S_F/B. As defined in Figure 4.1, N=3 and S_F/B= 20 ft / 41 ft = 0.488. From Figure 4.1, for N=3 and extrapolating slightly for ε = 0.358

C_{Dg} = 1.09, for S_F/B= 0.5
C_{Dg} = 1.03, for S_F/B= 0.33

Interpolating for S_F/B= 0.488,
C_{Dg} = 1.03 + (1.09-1.03)[(.488-.33)/(.5-.33)] = 1.086

Next, the gross area force coefficient C_{Dg} is converted into a force coefficient compatible with *ASCE 7* by means of Equation 4.2.

$C_f = C_{Dg}/\varepsilon$ = 1.086 / 0.358 = 3.03

The force coefficient could also have been determined directly using the alternate method of Appendix 4A. The alternate method is somewhat more time consuming, but it covers a wider range of solidity ratios and frame spacing ratios, and can also be used to determine more accurate load combinations.

The area of application of force A_e has already been determined per floor level during calculation of solidity ratio. The wind force transmitted to each floor level may now be found by Equation 4-1a, $F = q_z\, GC_f\, A_e$ as shown below. The total force on the structural frame and appurtenances F_s is 143.1 kips (639.9 kN), found by summing the forces at all levels in Table 5B.3.

TABLE 5B.3
Total Force - Structural Frame and Appurtenances - F_s

Floor Level	q_z (psf)	G	C_f	A_e (ft^2)	F (lbs)
0	36.0	0.85	3.03	125	11,590
1	42.4	0.85	3.03	454	49,577
2	48.8	0.85	3.03	402	50,525
3	51.7	0.85	3.03	236	31,424
				$F_s = \Sigma F =$	143,116

Note : To convert pounds force (lbs) to newtons (N) multiply F values in this table by 4.448.

These forces are due to wind acting on the frame only. Wind forces acting on the vessels, equipment and piping are computed in accordance with sections 4.1 and 4.3, and added at the levels where the items are located. The structure supports two horizontal vessels on floor level one and three horizontal exchangers on floor level two (see section 3.3 and Figure 3.2). All of these items are subject to transverse wind loads for the wind direction under consideration. The projected diameter is equal to the vessel diameter plus 1.5 ft (0.46 m) per 4.3.2.2. The force coefficient is determined from *ASCE 7* Table 6-7 (per 4.3.2.3). The vessels and exchangers were assumed to be moderately smooth. These properties are listed in Table 5B.4.

TABLE 5B.4
Equipment Properties

Equip	Floor Level	Vessel Dia. (ft)	Proj. Dia. D (ft)	Length B (ft)	Proj. Area A_f (ft^2)	B/D	C_f
V1	1	4	5.5	10	55	1.8	0.51
V2	1	16	17.5	32	560	1.8	0.51
E1	2	10	11.5	24	276	2.1	0.52
E2	2	10	11.5	24	276	2.1	0.52
E3	2	2	3.5	20	70	5.7	0.58

Note : To convert ft^2 to m^2 multiply A_f values in this table by 0.0929.

The wind load on a given vessel or exchanger is the sum of the wind loads on the supports, platforms, large connecting pipes, and the cylinder itself as summarized in Table 5B.5. This example problem only considers the load on the cylinder. For determining q_h, the height of each vessel was assumed to be the mid-floor elevation above the supporting floor level. A small improvement in accuracy could be obtained by using the actual top elevation for each piece of equipment.

TABLE 5B.5
Gross Wind Force - Equipment

Equipment	q_h (psf)	G	C_f	A_f (ft^2)	F (lbs)
V1	42.4	0.85	0.51	55	1011
V2	42.4	0.85	0.51	560	10293
E1	48.8	0.85	0.52	276	5953
E2	48.8	0.85	0.52	276	5953
E3	48.8	0.85	0.58	70	1684

Note : To convert pounds force (lbs) to newtons (N) multiply F values in this table by 4.448.

The projected area of piping and electrical was given as 20% of the projected vessel area. For example, on floor level one, the piping area is equal to 0.2(55 + 560) = 123 ft^2 (11.42 m^2) The wind load on piping per floor level is summarized in Table 5B.6.

TABLE 5B.6
Gross Wind Force - Piping per Level

Floor Level	q_h (psf)	G	C_f	A_f (ft^2)	F (lbs)
1	42.4	0.85	0.7	123	3103
2	48.8	0.85	0.7	124	3612

Note : To convert pounds force (lbs) to newtons (N) multiply F values in this table by 4.448.

The total equipment load per floor level is equal to the sum of all of the vessel, exchanger, and piping wind loads on that floor level. For purposes of determining the overall wind load on the structure, equipment and piping loads can be reduced to account for shielding effects (shielding of the equipment by upwind portions of the structure, shielding of portions of the structure by upwind equipment, and equipment to equipment shielding). Note that for purposes of designing individual vessels and supports, the loads should not be reduced.

Since the vessels and exchangers are in the wind shadow of the stairs, diagonal bracing, middle column and intermediate beams, it is appropriate to reduce the equipment load for shielding by the upwind frames. Per equation 4.4, the shielding factor η is given by

$$\eta = (1 - \varepsilon)^{(\kappa + 0.3)}$$

where $\varepsilon = 0.358$ as calculated previously. The term κ accounts for additional shielding that equipment provides to other equipment and to downwind frames.

If the additional shielding is considered, then κ must be determined for each floor level where this additional shielding exists. On floor level one, the smaller vessel V1 provides a limited amount of shielding for the larger vessel V2, and V2 provides significant shielding for the center column, bracing, and intermediate beam on the leeward frame. The volumetric solidity ratio κ is equal to the total volume of equipment on a floor level divided by the total volume of that floor level (per 4.2.6.2).

Equipment volume = $10 \pi (4^2)/4 + 32 \pi (16^2)/4 = 6560$ ft^3

Total volume = plan area x height = (40 x 40) x 28 = 44,800 ft^3

$\kappa = 6560 / 44,800 = 0.15$

$\eta = (1 - 0.358)^{(0.15 + 0.3)} = 0.82$

At the second floor level, exchanger E3 is windward of E1, but it is very small (2 ft (0.61 m) diameter), and exchangers E2 and E3 do not shield a significant portion of the leeward frame. Therefore, no additional shielding is appropriate; use $\kappa = 0$.

$$\eta = (1 - 0.358)^{(0 + 0.3)} = 0.88$$

Note that if the solidity of the upwind frames varies considerably from level to level, and if $\kappa > 0$ on any level, it would be appropriate to calculate an ε for each level supporting equipment rather than using a single value of ε for the overall structure.

Summing the loads on the vessels, exchangers, and piping, per level and applying the shielding factors yields the total wind load on equipment seen in Table 5B.7.

TABLE 5B.7
Total Wind Force - Equipment and Piping

Floor Level	Equipment and Piping Load	η	Reduced Load -lbs (kN)
1	V1+V2+piping = 1011+10293+3103 = 14407 lbs (64.1 kN)	0.82	11,814 (52.5)
2	E1+E2+E3+piping = 5953+5953+1684+3612 = 17203 lbs (76.5 kN)	0.88	15,138 (67.3)
	Total Equipment Load = Σ Reduced Loads =		26,952 (119.9)

5B.2 CROSSWIND FORCE CALCULATIONS

The next step is to repeat the analysis for the nominal wind direction normal to column line A (see Figures 3.2 and 3.4) - "non windward" frame. The member sizes are the same on this elevation except that the intermediate beams are W10's and

   Beams El 20 ft 0 in - W14
   Beams El 48 ft 0 in - W16
   Beams El 82 ft 0 in - W12

The gross area of the windward face includes the stair tower on the right hand side of the structure.

$$A_g = (83 \times 41) + (9 \times 49) = 3,844 \text{ ft}^2$$

The solid areas for the windward frame are given below. The stairs column in the table includes areas of stair column, struts, and handrails (See Table 5B.8).

TABLE 5B.8
Solid Area - A_S

Floor Level	Tributary Height (ft)	Solid Areas (ft^2)						
		Cols.	Beams	Int. Beams	Bracing	Hand Rails	Stairs	Total
0	0-10	30	0	0	19	0	24	73
1	10-34	72	46	8	35	40	44	245
2	34-65	93	53	41	36	40	36	299
3	65-83	51	40	0	16	40	0	147
		Total Solid Area of Windward Frame (ft^2) =						764

Note : To convert ft^2 to m^2 multiply A_S values in this table by 0.0929.

Since the solidity of neither the middle and leeward frames (column lines B and C, respectively) exceeds that of the windward frame, A_S is equal to the solid area of windward frame, yielding

$$\varepsilon = A_S / A_g = 764 \text{ ft}^2 / 3{,}844 \text{ ft}^2 = 0.199$$

The frame spacing ratio in this direction is S_F/B= 20 ft / 46 ft = 0.435. Since the width is not uniform (the stair tower stops at the second floor level), an average value of B was used. From Figure 4.1 for N=3 and ε = 0.199

C_{Dg} = 0.72, for S_F/B= 0.5
C_{Dg} = 0.71, for S_F/B= 0.33

Therefore use C_{Dg} = 0.716.

$$C_f = C_{Dg}/\varepsilon = 0.716 / 0.199 = 3.60$$

The wind forces per floor level are shown in Table 5B.9.

The wind direction is parallel to the axis of the vessels and exchangers (longitudinal wind). The vessels have rounded heads and the exchangers have flat heads. Force coefficients for this case are given in 4.3.2.4, and the wind loads are tabulated in Table 5B.10.

TABLE 5B.9
Total Force - Structural Frame and Appurtenances - F_s

Floor Level	q_z (psf)	G	C_f	A_e (ft^2)	F (lbs)
0	36.0	0.85	3.60	73	8,042
1	42.4	0.85	3.60	245	31,787
2	48.8	0.85	3.60	299	44,649
3	51.7	0.85	3.60	147	23,256
				$F_s = \Sigma F =$	107,734

Note : To convert pounds force (lbs) to newtons (N) multiply F values in this table by 4.448.

TABLE 5B.10
Gross Wind Force - Equipment

Equipment	q_h (psf)	G	C_f	A_f (ft^2)	F (lbs)
V1	42.4	0.85	0.5	13	226
V2	42.4	0.85	0.5	201	3624
E1	48.8	0.85	1.2	79	3910
E2	48.8	0.85	1.2	79	3910
E3	48.8	0.85	1.2	3	156

Note : To convert pounds force (lbs) to newtons (N) multiply F values in this table by 4.448.

The wind load in the piping and electrical is the same as calculated previously. For this wind direction, there is no significant shielding of the equipment by the windward frame and no equipment to equipment shielding, so no reduction is taken on equipment load. Summing the loads on the vessels, exchangers, and piping per level yields

Level 1: V1 + V2 + piping = 226 + 3624 + 3103 = 6,953 lbs (30.9 kN)

Level 2: E1 + E2 + E3 + piping = 3910 + 3910 + 156 + 3612 = 11,588 lbs (51.5 kN)

for a total equipment and piping load of F_E = 18,540 lbs (82.5 kN).

5B.3 Open Frame Example - Summary and Conclusion

The results thus far are summarized in the Table 5B.11. The load combinations for design are application of F_T in one direction simultaneously with 0.5 F_S in the other, per 4.2.6.1. These combinations are shown in Figure 5B.1(a).

TABLE 5B.11
Summary

	Wind - Direction 1	Wind - Direction 2
Wind Load on Structural Frame F_S	143 kips (636 kN)	108 kips (480 kN)
Wind Load on Equipment and Piping F_E	27 kips (120 kN)	19 kips (84.5 kN)
Total Wind Load on Structure F_T	170 kips (756 kN)	127 kips (565 kN)

Using the alternate method of Appendix 4A to solve this entire problem, the force coefficient for the first wind direction would be 2.92 instead of 3.03. The force coefficient for the second wind direction would be 3.44 instead of 3.60. Wind load on equipment is unaffected. The load combination factor on F_S for load case 1 would be 0.58 instead of 0.5, and the factor on F_S for load case 2 would be 0.41 instead of 0.5. The wind loads obtained using the alternate method are shown in Figures 5B.1(b).

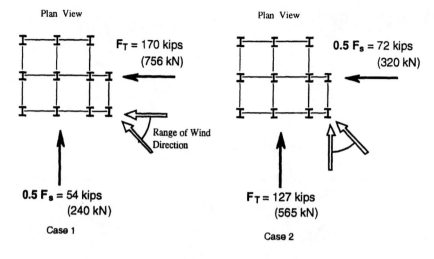

(a) Using method in guidelines

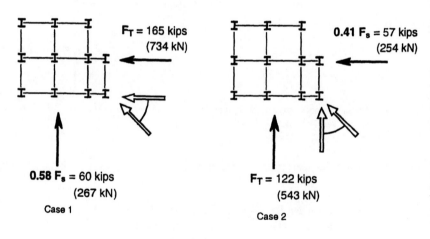

(b) Using method of Appendix 4A

Figure 5B.1 Design Load Cases for Open Frame Example

APPENDIX 5.C
PRESSURE VESSELS EXAMPLE

This section will demonstrate the application of the recommended guidelines for calculating wind loads on pressure vessels. Both the vertical and the horizontal vessels described in Section 3.4 and shown in Figures 3.5 and 3.6 will be considered.

5C.1 VERTICAL VESSEL

5C.1.1 Simplified Method - Rigid Vessel

Wind loads determined using Equation 4.1a; $F = q_z G C_f A_e$; are shown in Table 5C.1. The velocity pressure, q_z is determined from Table 5.1. Other terms in equation 4.1a are determined as follows:

$G = 0.85$ (*ASCE 7*, Section 6.6.1)

$h/D = 150/10 = 15$

Assume vessel is moderately smooth, therefore

$C_f = 0.6 + 8 \times 0.1/18 = 0.64$ (*ASCE 7*, Table 6-7)

Increased diameter to approximate appurtenances: (Section 4.3.1.2)

$D + 5$ ft. $= 10 + 5 = 15$ ft. or
$D + 3$ ft. $+$ dia. of largest pipe $= 10 + 3 + 1.5 = 14.5$ ft.
Largest controls, therefore, effective $D = 15$ ft. (4.57 m)

Therefore, $A_e = 15h$

Height increase to account for platform and vapor line above tangent line is 1 diameter, which is 10 ft. per section 4.3.1.2. Therefore total effective height of the structure is 160 ft. (48.77 m).

5-19

TABLE 5C.1
Simplified Method - Calculation of Base Shear

Ht. Above Ground	q_z(psf)	G	C_f	A_e (ft^2)	F (lbs)
0 - 15	36.0	0.85	0.64	225	4400
15 - 20	37.1	0.85	0.64	75	1500
20 - 40	41.5	0.85	0.64	300	6800
40 - 60	46.2	0.85	0.64	300	7500
60 - 80	49.6	0.85	0.64	300	8100
80 - 100	52.6	0.85	0.64	300	8600
100 - 120	54.5	0.85	0.64	300	8900
120 - 140	56.6	0.85	0.64	300	9200
140 - 160	58.3	0.85	0.64	300	9500

Total = 64,500 lbs. (287 kN)

5C.1.2 Detailed Method - Rigid Vessel

5C.1.2.1 Vessel + Miscellaneous

Wind loads determined using Equation 4.1a; $F = q_z G C_f A_e$; are shown in Table 5C.2. The velocity pressure, q_z is determined from Table 5.1. Other terms in equation 4.1a are determined as follows:

G = 0.85   (*ASCE 7*, Section 6.6.1)

h/D = 150/10 = 15

Assume vessel is moderately smooth, therefore

C_f = 0.6 + 8x0.1/18 = 0.64   (-*ASCE 7*, Table 6-7)

Increased diameter to approximate ladder, nozzles & piping 8" or smaller:

D + 1.5 ft. = 10 + 1.5 = 11.5 ft. (3.51 m)   (Section 4.3.1.3)

Therefore, $A_e = 11.5h$

TABLE 5C.2
Detailed Method - Vessel & Miscellaneous - Calculation of Base Shear

Ht. Above Ground	q_z(psf)	G	C_f	A_e (ft^2)	F (lbs)
0 - 15	36.0	0.85	0.64	172.5	3400
15 - 20	37.1	0.85	0.64	57.5	1200
20 - 40	41.5	0.85	0.64	230	5200
40 - 60	46.2	0.85	0.64	230	5800
60 - 80	49.6	0.85	0.64	230	6200
80 - 100	52.6	0.85	0.64	230	6600
100 - 120	54.5	0.85	0.64	230	6800
120 - 140	56.6	0.85	0.64	230	7100
140 - 150	58.0	0.85	0.64	115	3600

Total = 45,900 lbs. (204 kN)

5C.1.2.2 Pipe:

Wind loads determined using Equation 4.1a; $F = q_z G C_f A_e$; are shown in Table 5C.3. The velocity pressure, q_z is determined from Table 5.1. Other terms in equation 4.1a are determined as follows:

$G = 0.85$ (*ASCE 7*, Section 6.6.1)

$C_f = 0.7$ (Section 4.3.1.3)

Pipe dia. = 18" = 1.5 ft. (0.46 m)

Therefore, $A_e = 1.5h$ (except for the curved section above El. 150 which must be calculated separately - see Figure 3.5)

A_e above El. 150 = $3.14 \times 10/2 \times 1.5 = 24$ ft^2 (2.2 m^2) -   {Note - the pipe starts at El. 15.0 (see Figure 3.5)}

TABLE 5C.3
Detailed Method -Pipe - Calculation of Base Shear

Ht. Above Ground	q_z (psf)	G	C_f	A_e (ft^2)	F (lbs)
15 - 20	37.1	0.85	0.70	8	180
20 - 40	41.7	0.85	0.70	30	740
40 - 60	46.2	0.85	0.70	30	820
60 - 80	49.6	0.85	0.70	30	890
80 - 100	52.6	0.85	0.70	30	940
100 - 120	54.5	0.85	0.70	30	970
120 - 140	56.6	0.85	0.70	30	1010
140 - 150	58.0	0.85	0.70	15	520
150 - 155	58.5	0.85	0.70	24	840

Total = 6,900 lbs. (30.7 kN)

5C.1.2.3 Platforms (Refer to Figure 3.5)

The platform at El. 150 (just above the top of the vessel) is a square platform 12 ft. x 12 ft. in plan with handrail around the perimeter. It is assumed that the platform structural framing will be 8-inches deep (8"/12" = 0.70 sq. ft./lin. ft.).

Therefore, A_e (plat. fram.) =   0.7 x 12 =   8.4 ft.^{2}
A_e (front handrail) =   0.8 x 12 =   9.6 ft.^{2} (Table 4.1)
A_e (back handrail) =   0.8 x 12 =   9.6 ft.^{2} (Section 4.3.1.3)
  27.6 ft.^{2} (2.56 m^2)

q_z = 58.3 psf   (Table 5.1)

G = 0.85   (ASCE 7, Section 6.6.1)

C_f = 2.0   (Section 4.3.1.3)

F = $q_z G C_f A_e$ = 58.3 x 0.85 x 2.0 x 27.6 = 2700 lbs (12.0 kN)

The other platforms are circular and extend 3 ft. beyond the outside radius of the vessel. Therefore the radial distance (R) from the centerline of the vessel to the outside of the platform is 5 + 3 = 8 ft. (2.44 m). The angle (60^0, 90^0 and 180^0) shown on Figure 3.5 is the angle subtended by the ends of the platform as measured at the centerline of the vessel. Therefore, the projected length of the platform is calculated by the equation:

L = 2RSin(subtended angle/2)

Platform at El. 100 ft. - subtended angle = 60^0

Projected length = 2 x 8 x Sin (60/2) = 8.0 ft.

Assume platform framing is 6-in. deep (0.5 sq.ft./lin.ft.)

A_e (plat. fram.) = 0.5 x 8.0 = 4.0 ft.^{2}
A_e (handrail) = 0.8 x 8.0 = <u>6.4 ft.^{2}</u>   (Table 4.1)
                              10.4 ft.^{2} (0.97 m^2)

q_z = 53.4 psf   (Table 5.1)

G = 0.85   (*ASCE 7*, Section 6.6.1)

C_f = 2.0   (Section 4.3.1.3)

F = $q_z G C_f A_e$ = 53.4 x 0.85 x 2.0 x 10.4 = 940 lbs. (4.2 kN)

Platform at El. 75 ft. - subtended angle = 60^0

Projected length = 2 x 8 x Sin (60/2) = 8.0 ft.

Assume platform framing is 6-in. deep (0.5 sq.ft./lin.ft.)

A_e (plat. fram.) = 0.5 x 8.0 = 4.0 ft.^{2}
A_e (handrail) = 0.8 x 8.0 = <u>6.4 ft.^{2}</u>   (Table 4.1)
                              10.4 ft.^{2} (0.97 m^2)

q_z = 50.5 psf   (Table 5.1)

G = 0.85   (*ASCE 7*, Section 6.6.1)

C_f = 2.0   (Section 4.3.1.3)

F = $q_z G C_f A_e$ = 50.5 x 0.85 x 2.0 x 10.4 = 890 lbs. (4.0 kN)

Platform at El. 45 ft. - subtended angle = $90°$

Projected length = 2 x 8 x Sin (90/2) = 11.3 ft. Since the projected length is larger than the vessel diameter, the back handrail will be included (Section 4.3.1.3). Back handrail projected length = 3 ft. x Sin $45°$ x 2 sides = 4.2 ft.

Assume platform framing is 6-in. deep (0.5 sq.ft./lin.ft.)

A_e (plat. fram.) =	0.5 x 11.3	= 5.7 ft.^{2}	
A_e (front handrail) =	0.8 x 11.3	= 9.0 ft.^{2}	(Table 4.1)
A_e (back handrail) =	0.8 x 4.2	= <u>3.3 ft.^{2}</u>	(Section 4.3.1.3)
		18.0 ft^2. (1.67 m^2)	

q_z = 45.2 psf (Table 5.1)

G = 0.85 (*ASCE 7*, Section 6.6.1)

C_f = 2.0 (Section 4.3.1.3)

F = $q_z G C_f A_e$ = 45.2 x 0.85 x 2.0 x 18.0 = 1380 lbs. (6.2 kN)

Platform at El. 15 ft. - subtended angle = $180°$

Projected length = 2 x 8 x Sin (180/2) = 16 ft. Since the projected length is larger that the vessel diameter, the back handrail will be included (Section 4.3.1.3). Back handrail projected length = 3 ft. x Sin $90°$ x 2 sides = 6 ft.

Assume Platform framing is 6-in. deep (0.5 sq.ft./lin.ft.)

A_e (plat. fram.) =	0.5 x 16.0 =	8.0 ft.^{2}	
A_e (front handrail) =	0.8 x 16.0 =	12.8 ft.^{2}	(Table 4.1)
A_e (back handrail) =	0.8 x 6.0 =	<u>4.8 ft.^{2}</u>	(Section 4.3.1.3)
		25.6 ft.^{2} (2.38 m.^{2})	

q_z = 36.0 psf (Table 5.1)

G = 0.85 (*ASCE 7*, Section 6.6.1)

C_f = 2.0 (Section 4.3.1.3)

F = $q_z G C_f A_e$ = 36.0 x 0.85 x 2.0 x 25.6 = 1570 lbs. (7.0 kN)

Total shear on platforms = 2700 + 940 + 890 + 1380 + 1570 = 7480 lbs (33.3 kN)

Total shear on foundation:

Vessel & Miscellaneous	= 45,900 lbs
Pipe	= 6,900 lbs
Platforms	= 7,480 lbs
Total	= 60,280 lbs (268 kN)

5C.1.3 Analysis of Flexible Vessels

The only difference between loads resulting from the analysis of the vessel as "Flexible" vs "Rigid" is that "G_f" (the gust response factor for main wind-force resisting systems of flexible buildings and structures) is substituted for "G" in the rigid analysis. *ASCE* 7-95, Section 6.6 states that G_f shall be calculated by a rational analysis that incorporates the dynamic properties of the main wind-force resisting system. A method is provided in the commentary. However, various companies and individuals have developed their own "rational" analysis. If one chooses to compare the numbers between rigid and flexible in Table 3.3, one will find that the factors are not consistent between the various Design Practices. Comparison of procedures for flexible structures is beyond the scope of this report. Therefore, no attempt has been made to make the factors agree.

For the "Recommended" methods, the procedure outlined in *ASCE* 7-95 Commentary, section 6.6 was utilized to determine $G_f = 1.03$

- SIMPLIFIED (FLEXIBLE):

    Total Shear = (G_f / G) x 64,500
    = (1.03 / 0.85) x 64,500 =   78,200 lbs (348 kN)

- DETAILED (FLEXIBLE):

    Vessel + Misc.:
    Shear = (G_f/G) x 45,900 lbs
    = (1.03 / 0.85) x 45,900 =   55,600 lbs

    Pipe:
    Shear = (G_f / G) x 6900 lbs
    = (1.03 / 0.85) x 6900 =   8,400 lbs

Platforms:
Shear = (G_f/G) x 7480 lbs
      = (1.03 / 0.85) x 7480 =   9,100 lbs
Total   =   73,100 lbs (325 kN)

5C.2 HORIZONTAL VESSELS - (Refer to Figure 3.6)

5C.2.1 Transverse Wind (wind on the side of the vessel)

$F = q_z G C_f A_e$ (Equation 4.1a; A_e is defined in Section 4.2.5)

h = 20 ft. at platform level, q_z = 38.2 psf   (See Table 5.1)

5C.2.1.1 Vessel + Miscellaneous

G = 0.85   (ASCE 7, Section 6.6.1)

B / D = 50/12 = 4.2, assume vessel is moderately smooth, therefore

C_f = 0.5 + 3.2 x 0.1/6 = 0.55   (ASCE 7, Table 6-7)

Increased Diameter to approximate ladder, nozzles & piping 8" or smaller:
D + 1.5 ft. = 12 + 1.5 = 13.5 ft. (4.1 m)   (Section 4.3.2.2)

Therefore, A_e = 13.5 x 54 avg. = 729 ft.2 (67.7 m2)

Calculate Shear at Base:

$F = q_z G C_f A_e$ = 38.2 x 0.85 x 0.55 x 729 = 13,000 lbs (57.8 kN)

5C.2.1.2 Platform:

The platform is a rectangular platform 10 ft. x 30 ft. in plan with handrail around the perimeter. It is assumed that the platform structural framing will be 10-inches deep (10"/12" = 0.8 sq. ft./lin. ft.).

Therefore, A_e (plat. fram.)   = 0.8 x 30 = 24.0 ft.^{2}
          A_e (front handrail) = 0.8 x 30 = 24.0 ft.^{2} (Table 4.1)
          A_e (back handrail) = 0.8 x 30 = 24.0 ft.^{2} (Section 4.3.2.6)
                                              72.0 ft.^{2} (6.7 m^2)

q_z = 38.2 psf   (Table 5.1)

$G = 0.85$ (*ASCE 7*, Section 6.6.1)

$C_f = 2.0$ (Section 4.3.2.6)

$F = q_z G C_f A_e = 38.2 \times 0.85 \times 2.0 \times 72.0 = 4700$ lbs (20.9 kN)

5C.2.1.3 Supports

Steel saddle

$A_e = 0.5 \times 2.0 \times 2$ (supports) $= 2.0$ ft^2 (0.2 m^2)

$q_z = 38.2$ psf (Table 5.1)

$G = 0.85$ (*ASCE 7*, Section 6.6.1)

$C_f = 2.0$ (Section 4.3.2.7)

$F = q_z G C_f A_e = 38.2 \times 0.85 \times 2.0 \times 2.0 = 130$ lbs (0.6 kN)

Concrete support:

$A_e = 1.0 \times 4.0 \times 2$ (supports) $= 8.0$ ft^2 (0.7 m^2)

$q_z = 38.2$ psf (Table 5.1)

$G = 0.85$ (*ASCE 7*, Section 6.6.1)

$C_f = 1.3$ (Section 4.3.2.7)

$F = q_z G C_f A_e = 38.2 \times 0.85 \times 1.3 \times 8.0 = 340$ lbs (1.5 kN)

5C.2.1.4 Total Shear, Transverse Wind

Vessel & Miscellaneous	= 13,000 lbs
Platform	= 4,700 lbs
Supports (steel)	= 130 lbs
Supports (concrete)	= 340 lbs
Total	= 18,170 lbs. (80.8 kN)

5C.2.2 Longitudinal Wind (wind on the end of the vessel)

$F = q_z G C_f A_e$

h = 20 ft. at platform level, q_z = 38.2 psf (See Table 5.1)

5C.2.2.1 Vessel + Miscellaneous

G = 0.85 (*ASCE 7*, Section 6.6.1)

Elliptical head, therefore C_f = 0.5 (Section 4.3.2.4)

Increased diameter to approximate ladder, nozzles & piping 8" or smaller:

D + 1.5 ft. = 12 + 1.5 = 13.5 ft. (4.1 m) (Section 4.3.2.2)

Therefore, A_e = 3.14 x 13.5 x 13.5/4 = 143.1 ft.^{2} (13.3 m^2)

Calculate shear at base:

F = $q_z G C_f A_e$ = 38.2 x 0.85 x 0.5 x 143.1 = 2320 lbs (10.3 kN)

5C.2.2.2 Platform:

The platform is a rectangular platform 10 ft. x 30 ft. in plan with handrail around the perimeter. It is assumed that the platform structural framing will be 10-inches deep (10"/12" = 0.8 sq. ft./lin. ft.).

Therefore, A_e (plat. fram.)   = 0.8 x 10 = 8.0 ft.^{2}
          A_e (front handrail) = 0.8 x 10 = 8.0 ft.^{2} (Table 4.1)
          A_e (back handrail) = 0.8 x 10 = <u>8.0 ft.^{2}</u> (Section 4.3.2.6)
                                          24.0 ft.^{2} (2.2 m^2)

q_z = 38.2 psf (Table 5.1)

G = 0.85 (*ASCE 7*, Section 6.6.1)

C_f = 2.0 (Section 4.3.2.6)

F = $q_z G C_f A_e$ = 38.2 x 0.85 x 2.0 x 24.0 = 1560 lbs (6.9 kN)

5C.2.2.3 Supports

Steel saddle - 10 ft. wide x 3 ft. (avg.) high

A_e = 10.0 x 3.0 x 2 (supports) = 60.0 ft^2 (5.6 m^2)

$q_z = 38.2$ psf (Table 5.1)

$G = 0.85$ (*ASCE 7*, Section 6.6.1)

$C_f = 2.0$ (Section 4.3.2.7)

$F = q_z G C_f A_e = 38.2 \times 0.85 \times 2.0 \times 60.0 = 3900$ lbs (17.3 kN)

Concrete support - 11 ft. wide x 4 ft. high

$A_e = 11.0 \times 4.0 \times 2$ (supports) $= 88.0$ ft^2 (8.2 m^2)

$q_z = 38.2$ psf (Table 5.1)

$G = 0.85$ (ASCE 7, Section 6.6.1)

$C_f = 1.3$ (Section 4.3.2.7)

$F = q_z G C_f A_e = 38.2 \times 0.85 \times 1.3 \times 88.0 = 3710$ lbs (16.5 kN)

5C.2.2.4 Total Shear, Longitudinal Wind

Vessel & Miscellaneous	= 2,320 lbs
Platform	= 1,560 lbs
Supports (steel)	= 3,900 lbs
Supports (concrete)	= 3,710 lbs
Total	= 11,490 lbs. (51.1 kN)

CHAPTER 6
RESEARCH NEEDS

6.0 GENERAL

To date, limited research has been conducted (or at least reported in the open literature) specifically related to wind loads on the types of industrial structures addressed in this report. The standards process requires that provisions be based on published, peer-reviewed research findings. This fact, in part, explains why pipe racks and most types of open frame structures and vessels found in the process industries are not addressed in *ASCE 7*. The lack of guidance from wind loading codes and standards and the absence of specific information in the technical literature has forced practicing engineers to make many assumptions and extrapolations far beyond the intended scope of the code and standard provisions, extrapolations often based on experience and engineering judgment alone. Differences in experience and judgment are manifested in the variety of procedures used by different firms and the sometimes accompanying large variations in estimated loads, as documented in Chapters 2 and 3 of this report.

It is understandable that the significant differences mentioned above could occur, given that the true nature of wind-structure interaction is often highly nonlinear and nonintuitive. In these cases, extrapolation beyond the boundaries of what is understood is a sometimes risky, but nonetheless necessary, venture. The recommendations set forth in Chapter 4 are based as much as possible on the solid foundation provided by *ASCE 7*. Beyond that, guidelines were based on the limited research data available, the current state of engineering practice, and the combined experience and judgment of the members of the Wind Induced Forces Task Committee. At least as important as the provisions of the recommended guidelines is the material contained in the commentary sections, which whenever possible gives the basis, rationale, and limitations for the guidelines.

The foregoing discussion clearly points out the need for research on wind effects on industrial structures. Results of the comparative study and the investigation of available research data (or lack thereof) for the different structure types were used to identify and prioritize some of the most important questions which need to be answered.

In an overall sense, piperack structures exhibited the greatest variation in loads between design guides of different companies and the greatest uncertainty in developing a consensus method for analysis. The committee feels that piperack structures in general should have the highest priority in any research program. Some specific items of concern include: determining appropriate force coefficients for pipes in a piperack structure; investigation of shielding and effects of different sizes of pipes together on the same level; force coefficients and shielding behavior of cable trays in a piperack; and investigation of wind loads on a vertical row of pipes, with wind direction ranging from parallel to perpendicular to the row.

Open frame structures were the next greatest cause of concern, with base shears on the structure used in the comparative study varying from highest to lowest by a factor of two. Wind tunnel test results on models simulating the steel framing alone are available and were used in developing the guidelines presented in Chapter 4, but no data are available on the effects of equipment and piping within the steel frames. Some critical questions to be answered for open frame structures include: Is there an upper bound on the load, and if so, what is it? What are the effects of equipment and vessels within the structure? Does solid or open flooring affect the loads?

Although some vessel loads also varied significantly from company to company, vessels were given the lowest overall priority, perhaps because the general procedures for calculating these loads were fairly similar. All firms used force coefficients for the basic shapes from *ASCE* 7-88 Table 12, differing primarily in treatment of surface roughness and projected areas to use. Specific questions which need to be answered include investigating the effects of platforms, large vertical pipes and smaller obstructions such as nozzles, ladders and small pipes.

6.1 RESEARCH PRIORITIES

Discussions among the committee led to the following prioritized list of the most significant unanswered research questions as:

1. What are appropriate force coefficients and shielding effects for pipes in a piperack structure?

2. What are appropriate force coefficients and shielding effects for cable trays in a piperack structure?

3. Is there an of the upper bound for wind induced forces on open frame structures that are rectangular in shape? If so, how is it determined?

4. What are the effects of 3D bluff bodies (equipment and vessels) in an open frame structure?

5. What are the loads on vertical rows of pipes (force coefficients for wind directions from perpendicular to parallel to a row)?

6. What force coefficients should be used for different size, type and orientation of platforms on a vessel?

7. What are the wind load interactions between a large vertical pipe and a vertical vessel?

8. What are appropriate force coefficients for ladders, nozzles and small pipes on a vessel? Can these items be accounted for by a simple increase in vessel diameter as proposed in Chapter 4 recommendations?

9. Do flooring and interior framing significantly affect loads on open frame structures? If so, how?

10. What are the effects of irregular (nonrectangular) plan view open frame structures on the magnitude and application of the wind induced force?.

11. What is the wind environment in a large petrochemical or energy facility? Does it significantly affect the wind loads?

NOMENCLATURE

A_e = area of application of force for the portion of the structure height consistent with the velocity pressure q_z, in square feet;

A_f = area of other structures or components and cladding thereof projected on a plane normal to wind direction, in square feet;

A_g = gross area of windward frame, in square feet;

A_s = effective solid area of windward frame, in square feet;

B = frame width or horizontal vessel length, in feet;

C_{Dg} = force coefficient for a set of frames, applicable to gross (envelope) area of the frames;

C_f = force coefficient;

D = outside diameter of circular pipe or vessel, in feet;

D' = depth of protruding elements (ribs or spoilers), in feet;

F = design wind force, in pounds;

G = gust effect factor for main wind-force resisting systems of rigid structures, (*ASCE 7 -95*);

$\overline{G}$ = gust response factor for main wind-force resisting systems of flexible buildings and structures (*ASCE 7-93* and previous versions);

G_f = gust effect factor for main wind-force resisting systems of flexible buildings and structures (*ASCE 7-95*);

G_h = gust response factor for main wind-force resisting systems of rigid structures evaluated at height z = h (*ASCE 7-93* and previous versions);

h = height of structure or vessel, in feet;

I = importance factor;

K_z = velocity pressure exposure coefficient evaluated at height z;

K_{zt} = topographic factor;

N = number of frames;

q_z = velocity pressure evaluated at height z above ground, in pounds per square feet;

S_F = frame spacing, in feet;

V = basic wind speed, in miles per hour;

z = height above ground, in feet;

ε = ratio of effective solid area of windward frame to gross area of windward frame;

κ = volumetric solidity ratio for the floor level under consideration. Defined as the ratio of the sum of the volumes of all equipment, vessels, exchangers, etc. on a level to the gross volume of that level;

η = a reduction factor to account for shielding of equipment by the structure and shielding of equipment by equipment;

GLOSSARY

Open Frame Structure - Open frame structures support equipment and piping within an open structural frame unenclosed by siding or other shielding appurtenances.

Pipe Rack - An open frame structure whose primary purpose is the support of piping and cable trays. Pipe racks are normally 15 ft to 25 ft (45.7m to 76.2m) in width and have two (2) or more levels.

Pressure Vessel - A container usually cylindrical in shape containing of a gas and/or liquid under pressure.

REFERENCES

Willford/Allsop Willford, M. R., and Allsop, A. C., *"Design Guide for Wind Loads on Unclad Framed Building Structures During Construction: Supplement 3 to The Designer's Guide to Wind Loading of Building Structures,"* Building Research Establishment Report, Garston, UK, 1990.

Cook Cook, N. J., *"The Designer's Guide to Wind Loading of Building Structures Part 2: Static Structures,"* Butterworths, London, 1990.

Georgiou Georgiou, P. N., *"Wind Loads on Building Frames"* M.E.Sc. Thesis, University of Western Ontario, *Canada*, 1979.

Georgiou/Vickery/Church, Georgiou, P. N.; Vickery, B. J.; and Church, R, *"Wind Loading on Open Framed Structures,"* Program and Workshop Notes, CWOWE III: Third Canadian Workshop on Wind Engineering, Vancouver, April, 1981, VI.1 pp. 1-19.

Whitbread Whitbread, R. E., *"The Influence of Shielding on the Wind Forces Experienced by Arrays of Lattice Frames"* Wind Engineering: Proceedings of the Fifth International Conference on Wind Engineering (Fort Collins, Colorado, USA, July, 1979), J. E. Cermak, Ed., Pergamon Press, 1980, pp. 405-420.

Walshe Walshe, D. E., *"Measurements of Wind Force on a Model of a Power Station Boiler House at Various Stages of Erection,"* NPL Aero Report 1165, National Physical Laboratory, Aerodynamics Division, Teddington, UK, September, 1965.

ASCE Wind ASCE, *"Wind Forces on Structures"*, Transactions of the ASCE Vol 126, Pages 1124-1198, 1962.

Nadeem Nadeem, A, *"Wind Loads on Open Frame Structures"*, M.S. Thesis, Louisiana State University, Baton Rouge, Louisiana, 1995.

Nadeem/Levitan Nadeem, A, and Levitan, M.L., *"A Refined Method for Calculating Wind Loads on Open Frame Structures"*, Proceedings, Ninth International Conference on Wind Engineering (January 9-13; New Delhi, India), 1995.

ASCE 7 The 1995 edition of ASCE 7 and its predecessors. When noted by itself this reference is the 1995 version of this document. ASCE 7 is titled *"Minimum Design Loads for Buildings and other Structures"*. This document is the successor of *ANSI A58.1*.

ANSI A58.1 American National Standards Institute *ANSI A58.1 "Minimum Design Loads for Buildings and Other Structures"*, 1982.

INDEX

A

American National Standards Institute (ANSI): ANSI A58.1, current design practices 2-1, 2-2—2-3
American Society of Civil Engineers (ASCE): ASCE 7-88, current design practices 2-1, 2-2—2-3; ASCE 7-95, differences in 3-1—3-2; ASCE 7—95, guideline 4-1; ASCE 7, scope of 1-1—1-2
Appurtenances: base shear calculation 5-20—5-21; calculation considerations 2-6; current vessel designs 2-7—2-8; force coefficients example 5-11, 5-16; longitudinal wind calculations 5-28; solid areas 4-11—4-12; transverse wind calculations 5-26—5-27
Area of application of force (A_e) 4-12

B

Base shear: comparison on vessels 3-13—3-15, 3-17—3-18; horizontal vessels 5-26—5-29; transverse wind calculations 5-27; vertical vessel example 5-20; vertical vessels 5-19—5-26
Bent space 4-2

C

Cable trays: comparison criteria 3-2—3-3; force coefficients 4-3; tributary areas 4-3; wind force method comparison 3-4—3-5. See also Pipe racks
Calculations: along wind force 5-10—5-14, 5-17; base shear, horizontal vessels 5-26—5-29; base shear, vertical vessels 5-19—5-26; crosswind force calculations 5-14—5-17; for secondary axis load combinations 4-22—4-24
Classification of structure categories, changes in 3-1
Codes and specifications, current design practices 2-2—2-3, 2-5, 2-6—2-7
Components: base shear calculation 5-20—5-21; calculation considerations 2-6; current vessel designs 2-7—2-8; force coefficients example 5-11, 5-16; longitudinal wind calculations 5-28; solid areas 4-11—4-12; transverse wind calculations 5-26—5-27
Configurations: case studies 1 and 2 comparisons 5-18; design wind force case study 4-14; frame plan calculation 4-28; horizontal vessel comparison 3-16; load combinations example 4-29; open frame structure comparison 3-6—3-8; open structure framing plan 4-8; pipe load cases I-IV 3-3; pipe rack plan example 5-5; vertical vessel comparison 3-12; wind directions and secondary axis 4-25; wind load versus wind direction 4-25

D

Decking 4-11
Design guide, in surveys 2-1
Design practices: horizontal vessel wind force comparisons 3-17—3-18; open frame structure survey 2-5; open frame structure wind force comparisons 3-10—3-11; pipe

rack survey 2-2—2-3; pipe rack wind force comparisons 3-1—3-5; pressure vessel survey 2-7—2-8; vertical vessel wind force comparisons 3-13—3-15

Design wind force: case study 4-13—4-16; equation, example 5-1; equation, explained 1-1; frame load equation 4-5—4-6; main force resisting systems equation 4-1; variances in practice 2-1. *See also* Winds

E

Effective solid area (A_s) 4-12

Equations: area of application of force (A_e) 4-12; design wind force, F (1.1) 1-1, 4-1, 5-1; effective solid area (A_s) 4-12; for frame loads, F_s (4.1a) 4-5—4-6; framing ratio, C_f (4.2) 4-7; main wind force resisting system 4-4—4-5; shielding factor, (4.4) 4-15; solidity ratio, (4.3) 4-11

Equipment: considerations 4-14, 4-15; force coefficient example 5-11—5-12, 5-16

Exposure categories, velocity pressure 1-2, 5-1

Exposure coefficients: changes in 3-1; determining 5-1

F

Flexible structures: gust effect factors 4-1, 5-2; load analysis of 5-25; recommendations on 2-4, 2-6; wind load comparisons 3-13—3-15. *See also* Pressure vessels

Floors, solid areas 4-11—4-12

Force coefficients: along wind force calculations 5-10—5-14, 5-17; components and cladding 4-5—4-6; consideration 1-2; crosswind force calculations 5-14—5-17; defined 4-2; for frame sets 4-7—4-11; frame sets and secondary axis 4-21—4-22, 4-26—4-27; load combination example 4-29; maximum 4-7; open frame practices 2-5; pipe rack practices 2-2—2-3; pipes and cable trays 4-3; pressure vessel practices 2-7—2-8

Frame loads: design wind force equations 4-5—4-6; force coefficients for frame sets 4-7—4-11; framing ratio 4-7; solid area 4-11—4-12

Frame space ratio: calculation 5-10—5-11; crosswind force calculation 5-15; equation 4-7

G

Gross (envelope) area 4-7, 5-10

Gust effect factors: changes in ASCE 7-95 3-1—3-2; defined 4-1; determining 5-2; determining for open frame structures 5-9; values of 4-1

Gust response factors 1-2

H

Horizontal vessels: design comparisons 3-17—3-18; design practices 2-7—2-8; determine force coefficients 4-18—4-20; longitudinal winds 5-27—5-29; transverse winds 5-26—5-27. *See also* Pressure vessels

I

Importance factors: changes in ASCE 7-95 3-1; consideration 1-2; current design practices 2-1; determining 5-1

L

Load combinations: design load case study 4-13, 4-14; secondary axis calculations 4-22—4-24
Lower bound: pipe load cases I-IV 3-4. See also Wind loads

M

Main wind force resisting systems: considerations 4-4—4-5; example 5-11
Maximum probable loads 2-4, 3-4. See also Wind loads
Minimum probable loads 3-4. See also Wind loads

O

Open frame structures: along wind force calculations 5-10—5-14, 5-17; alternate method 4-21—4-29; comparison criteria 3-5—3-9; components 4-5—4-6; crosswind force calculations 5-14—5-17; current practice survey 2-5; defined B-1; design considerations 2-4; design load case studies 1 and 2 4-13, 4-14; example criteria 5-1—5-2, 5-9; frame sets 4-7—4-11; main wind force resisting systems 4-4—4-5; solidity equation 4-11—4-13; wind force method comparisons 3-9—3-11

P

Pipe racks: comparison criteria 3-2—3-3; comparison of wind force methods 3-4—3-5; in current practices survey 2-2—2-3; defined B-1; design considerations 2-1; example criteria 5-1—5-2, 5-3; guidelines for force coefficients 4-2; pipe and cable tray example 5-3—5-5, 5-7; structure example 5-4—5-7
Pipes: base shear calculation 5-21—5-22; force coefficients 4-3; solidity ratio and shielding example 5-13—5-14; tributary areas 4-2
Platforms: base shear calculation 5-22—5-25; longitudinal wind calculations 5-28
Pressure vessels: comparison criteria 3-11—3-12, 3-15—3-16; considerations 2-6; defined B-1; design comparisons 3-14—3-15, 3-17—3-18; design practices 2-7—2-8; determine force coefficients 4-16—4-20; and shear 3-13—3-15; vertical vessel example 5-19—5-25
Projected areas, calculations for 4-2

R

Rigid structures: gust effect factors 5-2; wind loads 3-13—3-15. See also Pressure vessels

S

Secondary axis: determining force coefficients 4-21—4-22; load combination calculations 4-22—4-24
Shielding 1-2; along wind example 5-13—5-14; current pipe rack practices 2-2—2-3; factor equation 4-15; factors 2-4; and oblique winds 4-13; wind load considerations 4-2
Solidity ratio: along wind force exam-

ple 5-10—5-11; crosswind force calculation 5-15; explained 4-11—4-12
Specifications and codes: current design practices 2-2—2-3, 2-6—2-7, 5-2
Spheres, force coefficients 4-19—4-20
Streamlining effect 4-12
Structure classification categories, changes in 3-1—3-2
Structures: determining gust effect factors 5-2; recommendtion on flexible 2-4, 2-6; rectangular 4-4; unclad framed 4-4; wind force considerations 3-11
Supports: longitudinal wind calculations 5-28—5-29; transverse wind calculations 5-27

T
Torsion 4-16
Tributary areas 1-2, 4-2—4-3

U
Upper bound: indirect shielding 2-4; on pipe loads 3-4. *See also* Wind loads

V
Velocity pressure: determining 4—1; example profile 5-2; exposure coefficients 3-1; formula 5-1; open frame structures profile 5-9
Vertical aspect ratios 4-7
Vertical vessels: base shear 5-20—5-25; design practices 2-7—2-8, 3-13–3-15; determine force coefficients 4-16—4-18; simplified method 5-19—5-20. *See also* Pressure vessels

Volumetric solidity ratio: determining 4-15; example 5-13—5-14

W
Wind loads: defined 4-13, 4-14; research considerations 6-1—6-3. *See also* Equations; Design practices; Design wind force
Winds: along wind force calculations 5-10—5-14, 5-17; angles of 4-7; crosswind force calculations 5-14-5-17; loads *versus* wind direction 4-25; longitudinal wind calculations 5-27—5-29; normal and quartering 4-21; oblique and shielding 4-13; transverse and longitudinal 3-17—3-18; transverse wind calculations 5-26—5-27. *See also* Design wind force
Wind speeds 1-2
Windward frame: solid area calculation 5-10

Design of Anchor Bolts in

Petrochemical Facilities

Prepared by the

Task Committee on Anchor Bolt Design

ASCE Petrochemical Energy Committee

This publication is one of five state-of-the-practice engineering reports produced, to date, by the ASCE Petrochemical Energy Committee. These engineering reports are intended to be a summary of current engineering knowledge and design practice, and present guidelines for the design of petrochemical facilities. They represent a consensus opinion of task committee members active in their development. These five ASCE engineering reports are:

1) *Design of Anchor Bolts in Petrochemical Facilities*

2) *Design of Blast Resistant Buildings in Petrochemical Facilities*

3) *Design of Secondary Containment in Petrochemical Facilities*

4) *Seismic Design and Evaluation of Petrochemical Facilities*

5) *Wind Loads on Petrochemical Facilities*

The ASCE Petrochemical Energy Committee was organized by A. K. Gupta in 1991 and initially chaired by Curley Turner. Under their leadership, the five task committees were formed. More recently, this committee has been chaired by Joseph A. Bohinsky, followed by Frank J. Hsiu.

Frank J. Hsiu	J. Marcell Hunt
Chevron Research and Technology Co.	Hudson Engineering Corporation
Chairman	Secretary
Joseph A. Bohinsky	Brown & Root, Inc.
William Bounds	Fluor Daniel, Inc.
Clay Flint	Bechtel, Inc.
John Geigel	Exxon Chemical Company
Ajaya K. Gupta	North Carolina State University
Magdy H. Hanna	Jacobs Engineering, Inc.
Steven R. Hemler	Eastman Chemical Company
Gayle S. Johnson	EQE International, Inc.
James A. Maple	J.A. Maple & Associates
Douglas J. Nyman	D.J. Nyman & Associates
Norman Rennalls	BASF Corporation
Curley Turner	Fluor Daniel, Inc.

ASCE Task Committee on Anchor Bolt Design

This report was prepared to provide guidance in the design of headed, cast-in-place anchor bolts for petrochemical facilities. Although the makeup of the committee and the writing of this document are directed at petrochemical facility design, these guidelines are applicable to similar design situations in other industries. This report should interest structural engineers with responsibility for designing foundations as well as operating company personnel responsible for establishing internal design and construction practices. The task committee was established to provide some uniformity in the criteria currently used in the petrochemical industry.

This report is intended to be a State-of-the-Practice set of guidelines. The guidelines are based on published information and actual design practices. A review of current practice, internal company standards, and published documents was conducted. The report includes a list of references to provide additional information.

In helping to create a consensus set of guidelines, a number of individuals provided valuable assistance.

J. Marcell Hunt	Randy Russ
Hudson Engineering Corporation	Jacobs Engineering Group, Inc.
Chairman	Secretary

Van Ai	Jacobs Engineering Group, Inc.
Ed Alsamsam	Sargent & Lundy Engineers
Don Boyd	Parsons SIP
Howard Edwards	Parsons SIP
John Falcon	Jacobs Engineering Group, Inc.
James Lee	Brown & Root, Inc.
Paul Morken	John Brown
Dale Mueller	Litwin Engineers & Constructors, Inc.
Sam Ramesh	Bechtel, Inc.
Alan Shive	Fluor Daniel, Inc.
Dan Stoppenhagen	Fluor Daniel, Inc.
Hikmat Zerbe	Brown & Root, Inc.

CONTENTS

Chapter 1: Introduction 1-1

    1.1 Background 1-1
    1.2 Objectives and Scope 1-1
    1.3 Current State of Research 1-2
    1.4 Future Codes and Procedures 1-2

Chapter 2: Materials 2-1

    2.1 Introduction 2-1
    2.2 Grades 2-1
    2.3 Fabrication and Welding 2-1
    2.4 Corrosion 2-1

Chapter 3: Design 3-1

    3.1 Introduction 3-1
    3.2 Petrochemical Anchorage Design 3-3
    3.3 Bolt Configuration and Dimensions 3-3
    3.4 Design Basis 3-4
    3.5 Distribution of Anchor Bolt Forces 3-7
    3.6 Checking Critical Modes of Failure 3-10
    3.7 Pier Design/Reinforcing 3-16

Chapter 4: Installation 4-1

    4.1 Introduction 4-1
    4.2 Sleeves 4-1
    4.3 Pretensioning 4-3
    4.4 Considerations for Vibratory Loads 4-7
    4.5 Considerations for Seismic Loads (Zones 3 & 4) 4-7

Nomenclature A-1
References B-1

CHAPTER 1
INTRODUCTION

1.1 BACKGROUND

Due to limited coverage of anchorages in the commonly used design codes, most petrochemical engineering firms and owner companies use an extrapolation, variation, or interpretation of *ACI 349, Appendix B, "Code Requirements for Nuclear Safety Related Concrete Structures,"* to design anchorages. Also, ACI Publication AB-81, *"Guide to the Design of Anchor Bolts and Other Steel Embedments,"* is frequently used as the basis for the design of anchorage systems for the petrochemical industry. The lack of a single, authoritative design standard has resulted in inconsistent design and details. This committee's work has been influenced by the need to develop a uniform anchor bolt design methodology that is acceptable throughout the petrochemical industry.

1.2 OBJECTIVES AND SCOPE

The objective of this committee report is to:

a. Evaluate current petrochemical industry anchor bolt design methods.

b. Gather information regarding proposed changes and new releases of design codes which contain procedures for anchor bolt design.

c. Provide the engineer with information and recommendations that supplement current codes for design of headed, cast-in-place anchor bolts.

The committee recognizes that several different types of anchorage systems are used in petrochemical facilities, but the most common type is a cast-in-place, headed bolt. Therefore, for this report, the committee limited its investigation and recommendations to the cast-in-place, headed bolt. This self-imposed limit should not be construed as an attempt to limit the importance of other types of anchorage systems, but it provided a means of focusing the committee's attention on the most commonly used device.

1.3 CURRENT STATE OF RESEARCH

Test results are limited for bolts that are in the upper range of sizes and embedment depths that are commonly used in petrochemical facilities. A majority of embedment depths that have been tested are less than 7.87 inches (200 mm) with very few, if any, greater than 21.65 inches (550 mm). Most bolt sizes that have been tested are less than 2 inches (50 mm) in diameter, and a majority of the tests have been performed on bolts that are 1 inch (25 mm) or less in diameter.

Very little testing has been done that accounts for the effect of reinforcing on anchor bolt capacity in shear, tension, or both. Limited testing has been done which attempts to identify edge distance or anchor spacing influences. Lee and Breen (1966) reported on results for 26 bolts and Hasselwander, Jirsa, Breen, and Lo (1977) published a report based on results for 35 bolts. Baily and Burdette also published a report in 1977 entitled *"Edge Effects on Anchorage to Concrete."*

Recently, Fuchs et al., published a code background paper, *"Concrete Capacity Design (CCD) Approach for Fastening to Concrete"*, which presented an approach different from the well-known provisions of *ACI 349*. Furche et al, performed pullout tests with headed studs placed near a free edge and recommended an empirical equation for calculating the failure load in their paper titled *"Lateral Blow-out Failure of Headed Studs Near a Free Edge"*.

1.4 FUTURE CODES AND DESIGN PROCEDURES

ACI Committee 318 is currently working to add a chapter on anchorages to a future issue or a supplement to *ACI 318*.

In 1991, ACI Committee 355 published, a *"State-of-the-Art Report on Anchorage to Concrete."* This is the first of a two-volume set which emphasizes behavior and does not include design methods and procedures. They are currently working on the second volume, which will be a design manual, for anchorages whose capacity is computed based on unreinforced failure cones.

The *ACI 349, Appendix B*, subcommittee is reviewing additional anchor bolt testing to cover situations not covered by present test results (bolts close to the edges of concrete, closely spaced bolts, etc.). They plan on rewriting *ACI 349, Appendix B*, after they review the results of these tests. The planned rewrite will be based upon the Concrete Capacity Design (CCD) method (See Section 3.1 of this report.) with modifications to account for the results from this additional testing.

Information from these various ACI committees indicates that changes will be made to some of the formulas and methodologies that are currently being used. The trend appears to be moving from the *ACI 349* method toward the CCD Method.

CHAPTER 2
MATERIALS

2.1 INTRODUCTION

This chapter provides basic materials and corrosion protection recommendations for anchor bolts. Selection of the proper grade, strength, weldability, and corrosion resistance must be considered so that the anchor bolt will perform as the engineer intends.

2.2 GRADES

Table 2.1 lists the ASTM specifications, yield strength, ultimate strength, and range of available diameters for materials commonly used for anchor bolts.

2.3 FABRICATION AND WELDING

Flux, slag, and weld-splatter deposits should be removed before galvanizing because the normal pickling process does not remove slag. Toe cracking at weldments around anchor plates is undetectable prior to galvanizing and is easily detected after galvanizing. A post-galvanizing inspection should be considered to detect these cracks.

Materials which have been quenched and tempered should not be welded or hot dip galvanized. High-strength materials should not be bent or welded since their strength and performance may be affected by bending or heating.

2.4 CORROSION

Anchor bolt service life requires that corrosion protection be an important design consideration. Anchor bolt material or coating system selection should provide a reliable and high-quality service life for an item that is relatively

Table 2.1: Common Materials for Anchor Bolts

	ASTM Specification	F_y, ksi (MPa)	F_{ut}, ksi (MPa)	Diameter Range, inches (mm)	Notes
Bolts and Studs	A193 Gr B6	85 (590)	110 (760)	to 4 (102)	NG
	A193 Gr B7	105 (720)	125 (860)	to 2 1/2 (64)	
		95 (660)	115 (790)	over 2 1/2 (64) to 4 (102)	
		75 (515)	125 (860)	over 4 (102) to 7 (180)	NG
	A307	———	60 (410)	to 4 (102)	WCE
	A325	92 (630)	120 (830)	1/2 (13) to 1 (25)	
		81 (560)	105 (720)	over 1 (25) to 1 1/2 (38)	Im
	A354 Gr BC	109 (750)	125 (860)	1/4 (6) to 2 1/2 (64)	
		99 (680)	115 (790)	over 2 1/2 (64) to 4 (102)	
	A354 Gr BD	130 (900)	150 (1030)	1/4 (6) to 2 1/2 (64)	SC,
		115 (790)	140 (970)	over 2 1/2 (64) to 4 (102)	NG
	A449	92 (630)	120 (830)	1/4 (6) to 1 (25)	
		81 (560)	105 (720)	over 1 (25) to 1 1/2 (38)	
		58 (400)	90 (620)	over 1 1/2 (38) to 3 (76)	
	A490	130 (900)	150 (1030)	1/2 (13) to 1 1/2 (38)	Im,NG
	A687	105 (720)	150 (1030)	5/8 (16) to 3 (76)	
Threaded Round Stock	A36	36 (250)	58 (400)	to 8 (200)	W
	A572 Gr 42	42 (290)	60 (410)	to 6 (150)	
	A572 Gr 50	50 (340)	65 (450)	to 2 (51)	
	A588	50 (340)	70 (480)	to 4 (102)	
		46 (320)	67 (460)	over 4 (102) to 5 (127)	
		42 (290)	63 (430)	over 5 (127) to 8 (200)	
	B21 Temper H02 Alloy UNS C46400 Alloy UNS C48200	27 (190)	60 (410)	1/2 (13) to 1 (25)	Br
		26 (180)	58 (400)	over 1 (25) to (51)	
		25 (170)	54 (370)	over 2 (51) to 3 (76)	
		22 (150)	54 (370)	over 3 (76)	
	B98 Temper H02, Alloy UNS C65100	20 (140)	55 (380)	1/2 (13) to 2 (51)	Br
	B98 Temper H02, Alloy UNS C65500	38 (260)	70 (480)	to 2 (51)	Br
	F1554 Gr 36	36 (250)	58 (400)	1/4 (6) to 4 (102)	W
	F1554 Gr 55	55 (380)	75 (520)	1/4 (6) to 4 (102)	WS1
	F1554 Gr 105	105 (720)	125 (860)	1/4 (6) to 3 (76)	

Notes:
- WCE Weldable if carbon equivalent $\leq 0.35\%$.
- Im   Impractical because of limited available length.
- SC   Susceptible to stress-corrosion cracking.
- NG   Galvanizing is not an option in the ASTM specification.
- W   Weldable.
- Br   Brass alloy.
- WS1 Weldable with Specification's Supplementary Requirements S1.

inaccessible for maintenance, repairs, or replacement due to corrosion. There are many factors and environmental exposure conditions that should be considered. The engineer may need to consult with materials specialists about corrosion protection during the anchor bolt material selection process.

2.4.1 Environmental Conditions

Anchor bolts near waterways and seashores require corrosion protection against wet-dry cycles and excessive salts. De-icing salts in runoff from areas with snow and ice can also be particularly corrosive to anchor bolts.

Anchor bolts located in controlled environments inside buildings should not require protection from atmospheric corrosion except for exposure to chemicals.

Anchor bolts encased in concrete should not require corrosion protection unless sulfates or chlorides are present in the concrete. Joints in concrete should be sealed to keep moisture from anchor bolts.

Galvanized and stainless materials can fail when subjected to corrosive chemicals such as acids or other industrial fumes. Such materials require additional coating systems.

Bare, uncoated, weathering steels should not be used where high concentrations of corrosive chemical or industrial fumes are present.

2.4.2 Codes and Specifications

2.4.2.1 American Concrete Institute (ACI)

ACI 318-89 requires that protection be provided from injurious amounts of oil, acids, alkalis, salts, organic materials, or other substances that may be deleterious to the concrete, the reinforcing, and the anchor bolts.

ASTM A767 and *A775* specify a zinc coating and an epoxy coating of steel bars for concrete reinforcing in highly corrosive environments.

The soluble chloride ion content in concrete is controlled by *ACI 318-89*, Section 4.3.1; also see the report by ACI Committee 222, *"Corrosion of Metals in Concrete."*

When external sources of chlorides are present, increased concrete cover and/or an epoxy coating should be provided for reinforcing bars, in accordance with *ACI 318-89*, Sections 4.3.2 and 7.7.5. Anchor bolts should be considered as an extension of the concrete, as noted in *ACI 318-89*, Section 7.7.6, which requires that exposed reinforcement, inserts and plates intended for bonding with future extensions be protected from corrosion.

2.4.2.2 American Institute of Steel Construction (AISC)

Anchor bolt corrosion and material selection is outside the scope of AISC specifications or codes. The *"Steel Design Guide Series, Volume 7"*, includes a chapter to assist in some of the practical aspects of design and application of anchor bolts.

The *"Steel Design Guide Series, Volume 1"* recommends that anchor bolts subjected to corrosive conditions be galvanized. If anchor bolts are galvanized, it is best to specify *ASTM A307* and *A36* material to avoid embrittlement that sometimes results when high-strength steels are galvanized.

Weathering steels may be used when anchor bolts are exposed to corrosive atmospheres, but it should be understood that they will rust and stain the foundation concrete if so exposed.

2.4.2.3 American Petroleum Institute (API)

API 650 states that when corrosion is a possibility, an additional thickness should be considered for anchors. It is recommended that their nominal diameter not be less than 1 inch (25 mm) and that a corrosion allowance of at least 1/4 inch (6 mm) on the diameter be provided.

API 620 recommends using stainless steel anchorage materials or providing a corrosion allowance when using carbon steels.

2.4.3 Corrosion Rates

There are substantial variations in corrosion rates even under relatively similar conditions. Corrosion rates that are cited or determined by technical sources can vary in actual service life. During the design of anchor bolt protection systems, materials and process engineers should be consulted to define the corrosive exposure elements and what material or coating system is best suited for protection.

2.4.4 Coatings

If anchor bolts are in an area where the environment is particularly corrosive or abrasive, special coatings are required. Protective coatings may be preferable to increasing the bolt diameter or the length of embedment for the anchor bolt assembly. Polyamide epoxies and urethanes for carbon steel anchor bolts provide protection against alternating wet-dry environments. Phenolic epoxy coatings provide protection for chemical and acid vapors or fumes which exist in some industrial atmospheres or environments.

2.4.4.1 Considerations for Coating

A corrosion allowance is not required for anchor bolts that are galvanized or coated. Anchor bolts that are not galvanized or coated should have a minimum corrosion allowance of 1/8 inch (3 mm) added to their diameter.

All types of protective coatings should be periodically inspected and maintained to prevented corrosion from reducing the design capacity of the anchor bolt assembly.

Anchor bolts should be kept free of accumulations of excess materials or debris that may contain or trap moisture around anchors. Concrete surfaces should be sloped to drain water. Avoid details which will create pockets, crevices, and faying surfaces that can collect and accumulate water, debris, and other damp materials around the anchor bolt.

Foundations located in areas with a high ground-water table are highly susceptible to corrosion. The diameter of anchor bolts exposed to surface drainage or ground water should be increased a minimum of 1/8 inch (3 mm) for corrosion protection unless a protective coating is provided.

The surfaces between base plates and the concrete and/or grout pads should be sealed to prevent the infiltration of corrosive elements. Cement-sand, dry-pack grout pads should be coated or sealed in areas with cyclic wet-dry environments.

The service life of a combined system of paint over galvanizing is substantially greater than the sum of lives of the individual coatings. Precautions must be taken to ensure adherence of the paint to the galvanized surface, which is smooth and does not permit mechanical locking of the coating film.

2.4.4.2 ASTM A153, Hot Dip Zinc

Galvanizing with a hot-dip zinc process provides a cost-effective and maintenance-free corrosion protection system for most general applications. Precautions against embrittlement should be taken by the designer, the fabricator, and the galvanizer in accordance with recommended practice in *ASTM A143*. A coating weight of 1 to 2.5 ounces/square foot (0.3-0.75 kg/m^2) is normal for the hot-dip process. A recommended thickness of 2.3 ounces/square foot (0.7 kg/m^2) is an average application requirement. A corrosion allowance should not be required or added to galvanized anchor bolts.

Steel material with a tensile strength less than approximately 125 ksi (860 MPa) can be hot-dip galvanized, if that is an option in the appropriate ASTM specification (See Table 2.1). Steel material with a greater strength should not be hot-dip galvanized. As the yield stress increases, the possibility of hydrogen embrittlement increases because hydrogen is absorbed into the steel during the pickling process. Blast cleaning rather than pickling should be used for alloy materials.

Electro-deposited zinc coating can be applied to 1 inch (25 mm) diameters and smaller. Material sizes larger than a 1 inch (25 mm) diameter can be coated with an inorganic zinc-rich paint or other coating system specifically selected for corrosion protection.

A wet storage stain (white rust) should be prevented. Wet stain is a "voluminous white or gray deposit." It can form when closely packed, newly galvanized items are stored or shipped in damp or poorly ventilated conditions. This should not be confused with the normal process of weathering of the galvanized coating.

2.4.4.3 Cold-Applied Zinc

A cold-applied, organic, zinc-rich compound primer or coating should be used for field touch-up of galvanized bolts that have areas damaged during shipment or erection. Commercial zinc products for touch-up are zinc-rich paint, zinc spraying, or brushed molten zinc. A touch up paint should have 94% zinc dust in the dry film and should be applied to a dry film thickness of 8 mils (0.20 mm), minimum. Refer to *ASTM A780* for additional information.

2.4.4.4 Insulation and Fireproofing

Anchor bolts encased in weathertight or cementious insulation or fireproofing required for equipment normally do not require corrosion protection. If conditions exist for moisture to collect under the insulation or fireproofing, the anchor bolts should be coated with a zinc-based primer or other coating similar to that to be used for the equipment. Two coats of primer, for a total dry film thickness of 3-4 mils (0.08 to 0.10 mm), should provide the necessary corrosion protection for this service.

2.4.5 Weathering Steel (ASTM A588)

Weathering steels develop a tight oxide coating that protects against corrosion of the substrate. In certain environments, they will provide a relatively maintenance-free application. The material will form a protective surface with loss of metal thickness of about 2 mils (0.05 mm). Steels conforming to *ASTM A588* will provide atmospheric corrosion resistance that is 4 to 6 times the corrosion resistance of ordinary carbon steels.

Bare weathering steel should not be submerged in water because this steel will not provide corrosion resistance greater than a black carbon steel in the same service. Bare weathering steel should not be exposed to recurrent wetting by salt water, spray, or fogs because the salt residue will cause accelerated corrosion.

Weathering steels may be painted or galvanized as readily as carbon steels, although their appearance may not be uniform as a result of the higher silicon content. If urethane foam or other fire-retardants are to be used to protect weathering steels, consideration should be given to the fact that they can be very corrosive when wet with water. If paint is a consideration, consult the foam supplier for a recommendation of the paint system that is compatible with their foam.

CHAPTER 3
DESIGN

3.1 INTRODUCTION

In the past, there has been a decided lack of guidance in building codes for the design of anchorages to concrete. As a result, engineers have used experience, knowledge of concrete behavior, and guidance from other design recommendations (such as *ACI 349, Appendix B*) for help in designing these anchorages. In the future, however, this will likely change as *ACI 318* is working to introduce a new section of code addressing this important area of design. The proposed new code language will also recommend a design method that is somewhat different from that currently used by most engineers in the petrochemical industry. This method, as described in the paper by *Fuchs et al.*, is currently considered the state-of-the-art in anchor bolt design in non reinforced concrete. However, it is not the current state of practice in the petrochemical industry, where due to the small concrete sections, the reinforcement is used for transfer of anchor bolt forces to concrete.

The method, called the Concrete Capacity Design (CCD) method, is similar in principle to the method used by *ACI 349, Appendix B*, but has the following differences:

a. The *ACI 349, Appendix B*, method uses a conical failure surface for both tensile and shear loading. The CCD method uses a pyramid.

b. The *ACI 349, Appendix B*, method uses a failure slope of 45°, as opposed to the 35° failure slope used by the CCD method.

c. The *ACI 349, Appendix B*, method uses formulas for tension and shear which are proportional to the square of the depth of embedment and edge distance, respectively. Instead of using an exponent of 2 in these equations, the CCD method uses an exponent of 1.5.

Other than the change from the cone to the pyramid model, which was invoked to simplify the computations, these changes are based on empirical evidence drawn from experimentation in the United States and Europe and on theoretical consideration of variables such as size effect. For more information on the basis for the proposed changes, the user is referred to the paper by *Fuchs, et al.* This paper details how testing has revealed that the CCD method is a more accurate predictor of concrete capacity for various anchorages, which has also been verified by probabilistic studies by *Klingner, et al.* Furthermore, the experimental evidence presented in the paper by *Fuchs et al.*, shows that, for certain conditions, the models used in *ACI 349, Appendix B,* can actually overpredict the anchorage capacity (when using a ϕ, strength reduction factor, of 1). Therefore, the anchor bolts at the deeper embedments found in petrochemical design can be undersized using this method. The reader should be cautioned, however, that the amount of testing done on anchor bolt arrangements, bolt sizes, and depths of embedments typically found in the petrochemical industry is extremely limited, and is constrained to bolt sizes and embedments found close to the minimum sizes used in the petrochemical industry. Therefore, it is difficult to draw definite conclusions about the accuracy of using either method for larger bolts and deeper embedments without further experimentation.

The effects of the changes on design, however, can be easily demonstrated by looking at the formulas. Based on the changes described above:

a. For tension, the *ACI 349, Appendix B,* method will give smaller capacity for shallow embedments and greater capacity at deeper embedments than the CCD method. This is attributed to the exponent difference (2 for the ACI method; 1.5 for the CCD method).

b. For both shear and tension and anchor bolts close to an edge or close to other bolts, the *ACI 349, Appendix B,* method allows smaller edge distances and bolt spacing before the capacity of the anchorage is reduced. This results from the change in angle of the failure surface. (45° ACI, 35° CCD)

c. For bolts close to an edge and subjected to shear, the *ACI 349, Appendix B,* method will give lower anchor capacities at close edge distances and larger anchor capacities at large edge distances than the CCD method, because of to the exponent difference on the edge distance.

d. For anchors close to a corner, the capacity of a bolt according to the *ACI 349, Appendix B,* method is higher because of severe reductions by the CCD method for biaxial load effects.

All of these factors point to the fact that design by the CCD method will generally produce more conservative designs for the bolt sizes and embedments typically found in petrochemical design. However, the paper by *Fuchs, et al.* also notes that the method was primarily developed for anchors in unreinforced concrete and that the use of reinforcement designed to engage failure cones/pyramids could substantially increase the load capacity of the anchorage. Early evaluation of the CCD method for typical examples in petrochemical design supports the observation of more conservative results and found that, without the use of reinforcing, this method would lead to unacceptably conservative concrete member sizes.

This report is intended to give guidance for design of anchorages found in the petrochemical industry. Therefore, based on the observations above, it will identify the critical steps in anchor bolt design and will make recommendations for providing reinforcing details to provide safe and economical reinforced concrete designs.

3.2 PETROCHEMICAL ANCHORAGE DESIGN

Design of foundations in petrochemical design often involves the anchorage of tall vessels and structures subjected to heavy wind and seismic forces, resulting in large diameter anchor bolts. To transfer the loads from the anchor bolts to the reinforced concrete foundation, the embedment length of these anchor bolts can sometimes become quite large, and it is not uncommon for these anchor bolts to control the depth of the foundation.

The size of the concrete members in which the anchorage is embedded is often limited by the available space, which is often severely restricted by piping and electrical conduit, as well as by other foundations and access requirements. Because of the size and configurations of anchorages used in the petrochemical industry, design decisions often involve different choices not found in other industries. The flow chart shown in Figure 3.2 shows the design path that an engineer typically follows when designing an anchorage to concrete using a headed bolt.

3.3 BOLT CONFIGURATION AND DIMENSIONS

3.3.1 Bolt Configuration

The bolt configuration consists of either a headed bolt or a steel rod with threads on each end. *ASTM A36* rods have one nut tack welded to rod at the end which is embedded in concrete. For high strength material which is not weldable (such as *ASTM A193*), two nuts are provided at the end embedded in concrete. The two nuts are jammed together before bolt installation to prevent loosening when the top nut is tightened.

ACI 349-90, Section B4.5.2 does not differentiate between *ASTM A36,* A307 and high strength bolts. For high strength bolt assemblies such as *ASTM A193*, due to the very high bearing stress at the nut, it is recommended that a standard washer, of material compatible with the threaded rod material, be used. See Section 3.6.4 for washer requirements at high strength bolts.

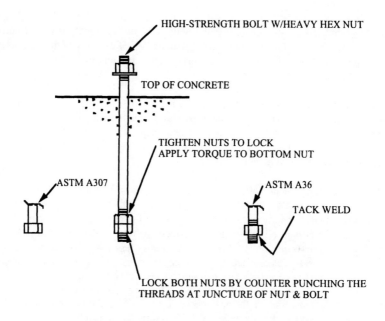

Figure 3.1: Anchor Bolt Head Configurations

3.3.2 Dimensions

The following minimum bolt dimensions are assumed in this section. The dimensions are insufficient for developing bolt loads in concrete and therefore pier reinforcement is required for load transfer. The dimensions are in conformance with the typical dimensions used in the petrochemical industry.

Embedment: Minimum embedment should be 12 bolt diameters
Minimum Edge Distance: 4 x bolt diameter for *ASTM A307* or *A36* bolts
6 x bolt diameter for high strength bolts
Minimum Bolt Spacing: 8 x bolt diameters

3.4 DESIGN BASIS

Depending upon the loads used for anchorage design and details of the anchorage, anchor bolt connections are classified as ductile or nonductile. For ductile connections, the embedment is proportioned using the ultimate capacity of the actual bolt. For nonductile connections, the embedment is proportioned using the factored design load.

A ductile connection can be defined as one that is controlled by yielding of steel elements (anchor bolt or reinforcement) with large deflections, redistribution of loads, and absorption of energy prior to any sudden loss of capacity of the anchorage resulting from a brittle failure of the concrete. *ACI 318* and other building codes favor ductile design.

As a minimum, anchorage design loads should be factored service loads, as required by *ACI 318*. However, there are valid reasons why the engineer may choose the design load to be the ultimate tensile capacity of the bolt. Reasons for making the anchor bolt, rather than the reinforcement, the "weak link" include easier detection and repair of damage from overload.

Sometimes client specifications dictate that design based on the ultimate capacity of bolts should be selected, but often the engineer must choose whether or not to use the ultimate tensile capacity of the bolt to determine the required embedment. This decision is an important one since it often affects the cost of the connections. The cost of connections using factored loads is generally less than those using ductile design as the design basis. Because of conservative bolt sizing by equipment manufacturers, corrosion allowances, and inherent conservatisms that result from the process of sizing a bolt by allowable stress and the concrete anchorage by ultimate strength, it is not uncommon for design based on the ultimate capacity of bolt to produce design forces on bolts which are more than twice the factored service loads.

The engineer should base the decision of design basis on client specifications, building code requirements, the nature of the applied loads, the consequence of failure, and the ability of the overall structural system to take advantage of the ductility of the anchorage.

3.4.1 Nature of Applied Loads

When peak loads are applied in a short-term or impulsive fashion, connections based on the ultimate tensile capacity of the bolt can enable a structural support to continue to carry loads until the short-term peak has passed.

Likewise, anchorage design should allow for the redistribution of loads and absorption of energy, as required in seismic or blast-resistant design. When the characteristics and magnitude of the load are unusually unpredictable, the anchorage design should be based on the ultimate tensile capacity of the bolt.

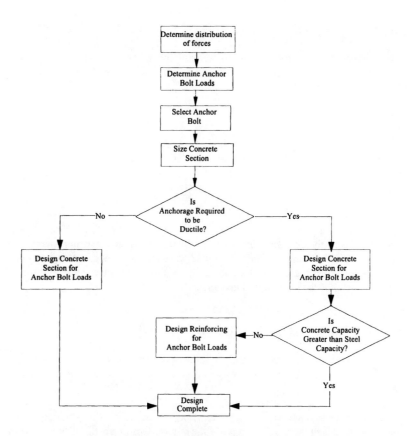

Figure 3.2: Headed Anchor Bolt Design Flowchart

3.4.2 Consequence of Failure

In some cases, the consequence of the failure of a single anchorage may be particularly undesirable. If, for instance, the failure of a single anchorage would lead to the collapse of a vessel or piping which contains highly flammable, toxic, or explosive materials (particularly polyfloric materials which explode when exposed to air), the engineer may want to base the anchorage design on the ultimate tensile capacity of the bolt. However, this decision depends on the characteristics of the structure, as described in the next section.

3.4.3 Ability of Overall Structural System to Take Advantage of Ductility

If yielding of a ductile connection produces a hinge in a structure which leads to or causes a brittle failure elsewhere in the structure, the benefits of this ductility are, at best, underutilized.

3.4.4 Conclusion

It is the opinion of this committee that transfer of anchor bolt load to reinforcement in the supporting member will produce a ductile design, provided the reinforcement is properly detailed to yield prior to concrete failure. Recommendations for detailing reinforcing steel are listed in Section 3.7 of this report.

Anchorage design should be approached as a global structural design issue, focusing more on the development of ductile load-resisting paths as opposed to the ductility of a single element. Once these load paths are developed, the engineer can then correctly assess the effect of a ductile connection and decide the requirements that should be imposed on an individual anchor.

3.5 DISTRIBUTION OF ANCHOR BOLT FORCES

Before an anchor bolt can be designed, the maximum force in the bolt must be determined. As stated above, this design basis should depend on the nature of the applied load.

An anchorage is ductile if the capacity of the concrete for each of the modes of failure is greater than the yield capacity of the bolt or reinforcement. If the bolt is to be the ductile element, then the design of the reinforced concrete elements depends only on the size of the bolt. The size of the bolt, however, can only be determined when the forces applied to the anchor group are distributed to the individual anchor elements (unless the bolt is sized by the equipment manufacturer).

If the anchorage is to be nonductile, the forces that are used to design the concrete elements are the forces that result when the factored loads are distributed to the anchor group. Sizing of the bolt requires that the distribution of forces under service loads be determined. Therefore, in both ductile and nonductile design, the forces applied must be distributed to the anchor elements. The assumptions regarding this distribution, however, can vary depending on the type of design chosen.

3.5.1 Distribution of Tension in Anchorages

Petrochemical structures supported by anchorages may be divided into two general categories. One is typified by a vertical vessel with anchor bolts around its periphery, and the other is typified by a structural column with anchor bolts clustered near the column.

3.5.1.1 Vessel Anchor Bolts

For a vertical vessel, anchor bolt maximum tension is commonly calculated by assuming an elastic distribution of forces and moments, which is based on the moment of inertia of the bolt group.

$$\text{Tension} = \frac{4M}{N \times BC} - \frac{W}{N} \qquad (3.1)$$

where:

M = maximum moment on vessel, kip-in (N-mm)
N = number of anchor bolts
BC = bolt circle diameter, in (mm)
W = minimum weight of vessel, kip (N)

Use of Equation 3.1 conservatively assumes that the moments are resisted only by the bolt group, and does not take into account the contributions of the base plate in resisting the moment, and does not consider strain compatibility between the concrete and steel elements which comprise the anchorage. Although a strain compatibility procedure is described in *Blodgett*, it is very complicated to perform by hand for most vessels with a ring baseplate and a circular anchor bolt arrangement. Properly including effects of flexibility of the vessel and its supporting skirt would further complicate analysis.

With a simplifying assumption of plane sections remaining plane and through the use of computers, however, it is possible to iterate to a solution which solves for the location of the neutral axis and determines the compressive forces under the baseplate ring and the tensile forces in the bolt. As stated previously, the design of anchorages for tall vessels and stacks can often control the design of the foundation. This procedure may, therefore, be worth undertaking when the above equation yields large anchor bolt sizes and embedments.

3.5.1.2 Structural Column Anchor Bolts

For a strength design, the design assumptions and general principles and requirements of the *ACI 318*, Chapter 10, should be used to determine capacity of an anchorage at a structural column. Anchor bolts should be sized as if they were reinforcement in a conventional concrete short beam-column.

For a structural column, anchor bolt tensions should be determined by a method which considers strain compatibility and equilibrium of forces and moments. One such method, based on reinforced concrete working stress design procedures, has been published by *Blodgett*.

Analysis of the anchorage as a reinforced concrete section should conform to the following guidelines, in addition to the requirements of *ACI 318*:

a. Anchor bolts should be considered as reinforcement, with area equal to anchor bolt tensile stress area. Tensile stress area is tabulated in the *AISC ASD Manual* and *AISC LRFD Manual* for bolt diameters up to 6 inches (150 mm).

b. If there is shear on the anchorage, it may be carried by anchor bolts in the compressive region of the section. If these bolts do not provide sufficient shear resistance, then analysis of the section should use a reduced anchor bolt area to account for reduced bolt tensile capacity. One expression for reduced anchor bolt area is

$$A_{eff} = A_t - \frac{V_u}{\phi \mu f_y} \qquad (3.2)$$

where:

A_{eff} = effective anchor bolt area for resisting tension, in² (mm²)
A_t = anchor bolt tensile stress area, in² (mm²)
V_u = ultimate shear per bolt, kip (kN)
ϕ = strength reduction factor = 0.85
μ = friction coefficient from *ACI 349, Appendix B*
  = 0.55 ....when the bottom of the base plate is raised above the concrete surface, as on a grout bed
  = 0.70 ....when the bottom of the base plate is on the concrete surface
  = 0.90 ....when the top of the base plate is at or below the concrete surface
f_y = anchor bolt yield stress, ksi (kPa)

c. The concrete cross-section should be the concrete area beneath the base plate.

For the analysis to be valid, two additional constraints must be present. First, the anchorage must develop the yield strength of the anchor bolt. Second, the base plate must be sufficiently thick (or stiffened) so that it does not form a mechanism with plastic hinges before the anchor bolts yield. One way to ensure adequate base plate thickness is to provide a thickness so that prying force, when calculated in accordance with the *AISC ASD Manual* and *AISC LRFD Manual* is insignificant.

3.5.2 Distribution of Anchor Bolt Shears

3.5.2.1 Means of Carrying Shear

In general, shear applied to an anchorage may be carried by one of the following mechanisms:

a. Friction due to net compression and to compression due to moment couple. (This is a nonductile mechanism.)

b. Anchor bolts.

c. Embedments such as shear keys. It is essential that grout completely fill a shear key slot, and use of grout holes in the base plate is suggested. When shear keys are used, they should be sized to carry the entire anchorage shear, because shear keys are stiffer than anchor bolts in shear.

3.5.2.2 Shear in Ductile Anchorage Groups

If shear capacity of anchor bolts not carrying tension exceeds total design shear, then anchor bolts carrying tension may be assumed to carry no shear.

If total design shear exceeds shear capacity of anchor bolts not carrying tension, then anchor bolts carrying tension should be assumed to carry the difference. Shear among tensile anchor bolts may be apportioned to minimize overall bolt requirements. This assumption is valid, since the plastic deformation capabilities of a ductile anchor allow the loads to redistribute in the most efficient way.

3.5.2.3 Shear in Nonductile Anchorage Groups

If friction capacity exceeds applied total design shear, then anchor bolts may be assumed to carry no shear.

If total design shear exceeds friction capacity, then all shear should be assumed to be carried by the anchor bolts. Each bolt should be assumed to carry shear in proportion to its cross-sectional area, neglecting bolts having shear capacity limited by edge distance.

3.6 CHECKING CRITICAL MODES OF FAILURE

Once the design force in the anchor bolt has been determined, the concrete and steel elements which comprise the anchorage must be designed to resist these forces.

The forces applied to the individual anchor must be safely transferred to the supporting member. This process involves checking the steel bolt for failure, as well as checking the various critical modes of failure for the concrete. For a bolt in tension, these modes of failure include:

* Bolt tensile failure
* Pullout (concrete tensile) failure
* Lateral bursting (blowout) failure
* Localized bearing failure
* Concrete splitting failure

At present, the anchor bolt design practice of the petrochemical industry is to size concrete elements by strength design methods and to size steel elements mostly by allowable stress design (ASD) methods.

3.6.1 Bolt Failure

Historically, the most common anchor bolt materials used in the petrochemical industry have been *ASTM A307* or *A36* for low-to-moderate strength requirements and *ASTM A193* Gr B7 for high-strength requirements. For these grades of bolts, many engineers have commonly used allowable stresses lower than those permitted by *AISC*. In many cases, these lower allowable stresses have been justified by the design engineer to account for uncertainty in analysis, to permit future additions to structures, to reflect the critical role that the anchor bolts play in the structural integrity of the structure, and to reflect the relatively low cost of anchor bolts when compared to the cost of failure. In other cases, engineers have simply been obligated to comply with owners' specifications which have specified lower allowable loads for the same basic reasons. The following table compares allowable loads specified by an owner, using tensile stress area, with those permitted by *AISC ASD* for a 2 inch (51 mm) diameter bolt.

Table 3.1: Comparison of allowable bolt loads for 2 inch diameter A36 and A193 Grade B7 bolts

Bolt Material	Owner allowable tension kips (kN)	*AISC ASD* tensile capacity kips (kN)	Owner allowable shear kips (kN)	*AISC ASD* shear capacity kips (kN)
A36	48 (10.7)	60 (13.3)	25 (5.5)	41 (9.1)
A193 Gr B7	100 (22.2)	130 (28.9)	26 (5.8)	86 (19.1)

3.6.1.1 Combined Tension and Shear

a. Ultimate Strength Design

For bolts subject to both tension and shear, the following conditions should be met:

$$\left(\frac{T_u}{\phi_1 P_n}\right)^k + \left(\frac{V_{ua}}{\phi_2 V_n}\right)^k \leq 1 \tag{3.3}$$

where:

T_u = factored tensile load per bolt, kips (kN)
ϕ_1 = 0.90, strength reduction factor for tension load
P_n = nominal tensile capacity of bolt, kips (kN)
V_{ua} = factored shear force per bolt, kips (kN)
ϕ_2 = 0.85, strength reduction factor for shear load

Required values of the exponent 'k' differ for various codes. *ACI 349, Appendix B*, states that bolt areas required for tension and for shear shall be additive, implying k = 1.0. The *Uniform Building Code* uses k = 5/3. The *AISC ASD* and *AISC LRFD* use an elliptical interaction curve, which is dependent on the bolt material for determining the strength of bolts subject to combined tensions and shear.

A 'k' value of 1, in the opinion of this committee, is very conservative. As the interaction equation is being applied for the steel stresses only, a 'k' value of 2 should be used.

$$\phi_1 P_n = \phi_1 F_t A_t \tag{3.4}$$

where:

F_t = the smaller of 1.0 times bolt yield stress or 0.9 times bolt tensile strength
A_t = bolt tensile stress area, in² (mm²)

$$\phi_2 V_n = \phi_2 \mu f_y A_t \tag{3.5}$$

where:

μ = friction coefficient from *ACI 349, Appendix B*
 = 0.55 ....when the bottom of the base plate is raised above the concrete surface, as on a grout bed
 = 0.70 ....when the bottom of the base plate is on the concrete surface
 = 0.90 ....when the top of the base plate is at or below the concrete surface

f_y = anchor bolt yield stress, ksi (kPa)
A_t = bolt tensile stress area, in^2 (mm^2)

b. Allowable stress design

The *AISC ASD* provides allowable stress design requirements for *ASTM A307* (valid for *A36*), *A325*, *A449*, and *A490* bolts. The expressions in the specifications which use stresses based on bolt nominal area, are applicable for tension alone, shear alone, and combined tension and shear interaction.

For threaded fasteners, per the *AISC ASD*, the allowable tensile stress is 0.33F_u and allowable shear stress is 0.22 F_u (threads excluded from the shear plane) or 0.17 F_u (threads included in shear plane). The *AISC ASD* does not specify interaction expressions except for the steels mentioned above. For other materials a linear interaction expression in the form given below would be conservative:

$$\left(\frac{f_t}{F_t}\right)^k + \left(\frac{f_v}{F_v}\right)^k \leq 1.0 \tag{3.6}$$

where:

f_t = calculated tensile stress, ksi (kPa)
f_v = calculated shear stress, ksi (kPa)
F_t = allowable tensile stress, ksi (kPa)
F_v = allowable shear stress, ksi (kPa)

A 'k' value of 1, in the opinion of this committee is conservative and a 'k' value of 2 is recommended.

3.6.2 Pullout Failure

A concrete pullout failure occurs when the tensile forces on the anchor bolts produce principal tensile stresses which exceed the capacity of the concrete along a failure surface projecting from the bolt head upward at an angle toward the direction of the applied load. This failure surface is what is commonly called the "shear cone" that resembles the model used by *ACI 349, Appendix B*.

The choice of whether to use the cone model (*ACI 349, Appendix B*) or the pyramid model (CCD method) is not significant if reinforcement is going to be used to transfer the design force, as will typically be the case for petrochemical anchorage designs. If reinforcement is not required to transfer the design loads or if no reinforcement crosses the failure surface, it is recommended that the engineer use the CCD method as described in the paper by *Fuchs, et al*. Experimental evidence presented in this paper indicates that the cone model and equations given by *ACI 349, Appendix B*, may overpredict the capacity of an anchor in unreinforced concrete.

Section 3.6 of this document provides recommendations for detailing reinforcing steel to transfer tensile forces and prevent pullout failure.

3.6.3 Lateral Bursting Failure

Lateral bursting occurs when a deeply embedded anchor is located too close to the edge of the concrete, resulting in directional differences in restraint stiffness around the anchor bolt head and a corresponding lateral strain concentration on the side of the free edge. The result is a cone blowout failure that propagates from the head of the bolt to the edge of the concrete.

The minimum edge distance required to prevent lateral bursting is discussed in Section 3.7.2.2.

3.6.4 Localized Bearing Failure

3.6.4.1 Background

Traditional design procedures rely upon bond strength to use J-shaped or L-shaped bolts in smaller sizes and use a bolt with a bearing plate in the larger sizes. The plate was sized based on concrete bearing strength. In *ACI 349, Appendix B* (1980), for instance, bearing strength for small loaded areas is taken as:

$$\phi B_n = (2)(0.85)\phi f_c^{'} \tag{3.7}$$

where:

ϕ = 0.7, strength reduction factor
B_n = Nominal bearing capacity, ksi (kPa)
f'_c = 28 day compressive strength, ksi (kPa)

This calculation gives a fairly large plate size. For example, for a 1-1/2 inch (38 mm) bolt, with f_y = 36 ksi (250 MPa) and f'_c = 3 ksi (21 MPa), the required plate size is 3.82 inches (97 mm) square, normally rounded up to a convenient size.

In *ACI 349-82*, a drastic revision occurred. The Code said that bearing could be ignored, providing that bearing area of the anchor bolt head is at least 1.5 times the tensile stress area. The commentary makes it clear that all normal bolt heads are included. The implied bearing strength (ignoring differences in ϕ factors) is 2/3 f_y, which for f_y = 36 ksi (250 MPa) and f'_c = 3 ksi (21 MPa) is 8 f'_c.

ACI 349-90 permits ignoring bearing under ordinary bolt heads of *A307, A325,* and *A490* material. Anchor heads for other materials are also acceptable provided they have at least the bearing area as above and a thickness equal to 1.0 times the distance from the edge of the anchor head to the "face of the tensile stress component."

ACI 349 states that if sufficient edge distance is provided to "develop the necessary confining pressure," crushing of the concrete under the bolt head will never occur.

3.6.4.2 Recommendations

Based on the latest research, the concrete bearing strength under the bolt head or anchor plate can reach a maximum of

$$0.75(11)f'c(A_{head} - A_{nominal}) \tag{3.8}$$

where 0.75 is ϕ, the strength reduction factor. Bond strength is assumed to be zero. Equate the concrete pullout strength with the expression for fastener strength:

$$(0.9)(A_{tensile})f_y = 0.75(11)f'_c(A_{head} - A_{nominal}) \tag{3.9}$$

(This equation is used with factored loads; when nominal loads are used, the allowable bearing can be taken as 4 f'_c.)

3-15

Then:

$$\frac{A_{head} - A_{nominal}}{A_{tensile}} = 0.11 \frac{f_y}{f'_c} \tag{3.10}$$

where:

$A_{tensile}$ = tensile stress area of bolt, in² (mm²)
A_{head} = area of bolt head, in² (mm²)
$A_{nominal}$ = nominal area of bolt, in² (mm²)

For f_y = 36 ksi (250 MPa) and f'_c = 3 ksi (21 MPa) the right hand side of the above equation has a value of 1.3. Since the ratio of head size to bolt size is greater than this for all available sizes of anchor bolts covered by this report, one can conclude that no special anchor plates are needed when using *A36* bolts. For higher-strength bolts, Equation 3.10 can be used to determine whether anchor plates are needed.

3.6.5 Concrete Splitting Failure

Splitting failures are caused by splitting of the structural member before failure of the anchor bolt. Splitting transverse to the tensile force can occur between anchor bolts in multiple bolt anchorages where the spacing between bolts is less than the bolt embedment depth. Also, a splitting failure can occur transverse to the tension force if a plane of closely spaced reinforcing steel is located near the embedded end of the anchor bolts.

If the bolts are spaced far enough apart so that (1) either concrete tensile stress area as recommended by *ACI 349* is provided for each bolt or properly developed reinforcing to transfer bolt load to the concrete is provided and (2) if a plane of closely spaced reinforcing steel does not exist near the embedded end of the anchor bolts, then a splitting failure should not occur.

3.7 PIER DESIGN/REINFORCING

The anchor bolt/reinforcement system is designed assuming that the anchor bolt tension and shear forces are resisted by the pier vertical reinforcement and ties respectively. Pier reinforcement is used to resist the anchor bolt tension and shear loads as it is generally not possible to provide the edge distances and bolt spacings required to carry anchorage loads in pier concrete alone.

It is always good detailing practice to include additional reinforcement near the head of the anchor bolt, particularly when the tensile force on the bolt is high. Ties in this area help restrain any unbalanced lateral forces which may result when the concentrated force at the bolt head is transferred to adjacent reinforcement.

3.7.1 Background

The commentary on Appendix B of *ACI 349-90* recommends use of inverted hairpin reinforcement, edge angles attached with reinforcing, and helical reinforcement to resist tension, shear, and lateral bursting, respectively. Although these provide a valid option from an engineering point of view, their use may cause construction difficulties due to congestion of reinforcement.

Therefore, the preferred method for load transfer is as follows:

Tension force ....transfer tension through vertical pier reinforcement.
Shear force ....transfer shear through ties in the pier.
Lateral bursting force.............provide minimum edge distance or ties at bolt head.

3.7.2 Transfer of Tension Force

Tension force in anchor bolts induces tensile stresses in concrete due to direct tensile load transfer and lateral bursting forces at the anchor bolt head. A recommended arrangement of reinforcement for resisting concrete tensile stress in piers of square, rectangular, and octagonal cross-section is shown in Figures 3.3 and 3.4.

3.7.2.1 Tension

Vertical pier reinforcement intercepts potential crack planes adjacent to the bolt head. The reinforcement should be developed on either side of the potential crack plane. Equation 3.11 is used for calculating the area of steel required.

$$A_{st} = \frac{T_u}{\phi F_y} \tag{3.11}$$

where:

A_{st} = the area of vertical pier reinforcement per bolt, in² (mm²)
T_u = factored (ACI 318-89, Chapter 9) tensile load per bolt, kips (kN)
F_y = minimum specified yield strength of reinforcement steel, ksi (kPa)
ϕ = 0.90, strength reduction factor (ACI 318-89, Chapter 9)

To be considered effective, the distance of the reinforcement from the anchor bolt head should not exceed the lesser of one-fifth of L_d or 6 inches, where L_d is the embedment length of the anchor bolt. There is no test validation for the one-third L_d space requirement; however, it has been specified in *ACI 349* and has been adopted in this section as a good practice.

In order to limit the embedment length of a bolt, a larger number of smaller-size bars is preferred over fewer, larger-size bars. In larger foundations, such as an octagon, two concentric layers of vertical reinforcement may be provided, as shown in Figure 3.4, if required, to transfer the anchor bolt tensile load.

The arrangement of reinforcement should take into consideration the minimum clearances required for placing and vibrating of concrete, minimum bar spacing required by *ACI 318*, and the need for adequate room below the bolt head or nut to ensure there is sufficiently compacted concrete.

The area of vertical pier reinforcement calculated using Equation 3.11 is not to be considered as additive to the reinforcement required strictly for resisting the moment and tension in sections of the pier. The calculated area of steel required for resisting the external loads applied to the pier should be compared with the area of steel required for resisting the tension in the anchor bolts. The area of vertical pier steel provided should equal or exceed the area of steel required for resisting the anchor bolt tension.

3.7.2.2 Lateral Bursting Force

Studies indicate that the minimum edge distance required for preventing lateral bursting of concrete at the bolt head, for *ASTM A307* or *A36* bolts and high strength bolts with specified minimum tensile strength equal to or less than 146 ksi (1009 MPa), is 4 times bolt diameter and 6 times bolt diameter, respectively. Therefore, the minimum edge distance, based on the bolt material, should be provided for all new bolt installations. In addition, for high strength bolts two sets of #3 (10 mm) ties at 3 inches (75 mm) spacing should be provided at the bolt head location. See Figures 3.3 and 3.4 for arrangement of ties at top and bottom of bolts in piers. The load transfer method outlined in this section is an extension of the requirements listed in the *ACI 318* and *349* concrete codes and the *ACI 355* report. Independent test to verify the proposed method have not been performed.

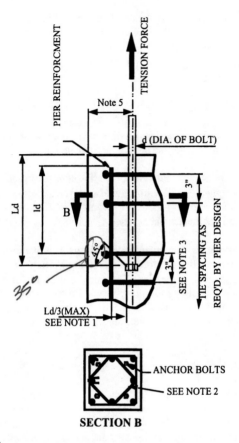

SECTION B

NOTES:
1) TO BE CONSIDERED EFFECTIVE FOR RESISTING BOLT TENSION, THE MAXIMUM DISTANCE FROM ANCHOR HEAD TO THE REINFORCEMENT SHALL BE Ld/3.

2) INTERIOR TIES TO BE PROVIDED IF REQUIRED PER ACI 318.

3) A MINIMUM OF 2 SETS OF TIES AT 3 INCH SPACING, CENTERED AT BOLT HEAD LOCATION, FOR HIGH-STRENGTH BOLTS ONLY. (SEE SECTION 3.7.2.2)

4) ld = DEVELOPMENT LENGTH OF PIER REINFORCMENT.
   Ld = EMBEDMENT LENGTH OF ANCHOR BOLT

5) 4d OR 4-1/2" MIN FOR ASTM A307/A36 BOLTS. 6 d OR 4-1/2" MIN FOR HIGH STRENGTH BOLTS (SEE SECTION 3.7.2.2)

Figure 3.3: Reinforcement for Resisting Bolt Tension in Square and Rectangular Pedestals

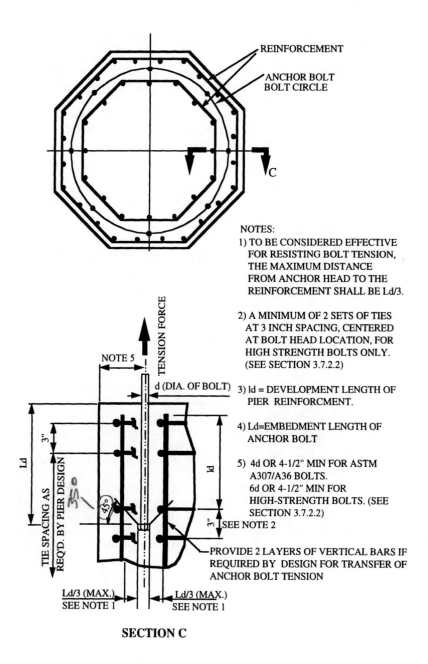

Figure 3.4: Reinforcement for Resisting Bolt Tension in Octagons

3.7.3 Transfer of Shear Force

Failure due to inadequate edge distance consists of breakout of a half prism or cone of concrete from centerline of bolt to face of pier in the direction of the shear force. If multiple bolts exist, then the failure prism/cones overlap. To reinforce the concrete in the failure plane, 2 sets of ties are provided at the top of piers. The first set is located at a maximum distance of 2 inches (50 mm) from top of concrete. The spacing between the two sets of ties should be 3 inches (75 mm).

Arrangement of ties to resist shear force in square and rectangular piers is shown in Figure 3.5. Transfer of shear force in octagons is generally not a problem and hence a detail has not been developed. Equation 3.12 is used to calculate the area of steel required to resist the shear force (factored load).

$$A_{sv} = \frac{V_u}{\phi F_y n} \qquad (3.12)$$

where:

A_{sv} = area of reinforcement required. Area of one leg of tie, in² (mm²).
V_u = factored shear force resisted by anchor bolt(s), kips (kN)
F_y = minimum specified yield strength of reinforcement steel, ksi (kPa)
n = number of legs in the top 2 sets of ties resisting the shear force
In Figure 3.5 the failure plane only intersects the top tie, therefore in this situation for Section A n = 1, for Section B n = 2.
ϕ = 0.85, strength reduction factor

The arrangement of reinforcement should consider the minimum clearances required for placing and vibrating of concrete and minimum bar spacing required by *ACI 318*. For large shear forces, shear lugs should be provided for transfer of loads. Also, in low seismic zones, shear may be transferred by friction between the base plate and top of pier, and anchor bolts may be used for transfer of tension only. The load transfer method outlined in this section is an extension of the requirements listed in *ACI 318, 349,* and *355.1*. Independent test to verify the proposed method have not been performed.

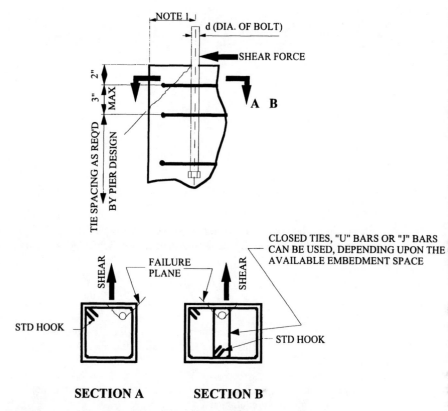

SHEAR CAPACITY OF TIES IS BASED ON NUMBER OF LEGS THAT INTERCEPT FAILURE PLANE IN PLAN AND ELEVATION. DEVELOPMENT LENGTH FOR TIE LEG MUST BE PROVIDED IN ORDER FOR TIE TO BE FULLY EFFECTIVE.

NOTES:

1) 4d OR 4 1/2" MIN FOR ASTM A307/A36 BOLTS. 6d OR 4 1/2" MIN FOR HIGH-STRENGTH BOLTS. (SEE SECTION 3.7.3.2).

Figure 3.5: Reinforcement for Resisting Bolt Shear in Square and Rectangular Pedestals

CHAPTER 4
INSTALLATION

4.1 INTRODUCTION

This chapter provides basic information regarding installation of anchor bolts with regard to sleeves, pretensioning, and considerations of vibratory or seismic loads.

4.2 SLEEVES

Sleeves are used with anchor bolts when a small movement of the bolt is desired after the bolt is set in concrete. This is generally required for one of the following two conditions:

a. When precise alignment of anchor bolts is required during installation of structural columns and/or equipment.

b. When anchor bolts are to be pretensioned in order to maintain the bolt under continuous tensile stresses during load reversal generated by high-pressure piping anchors, vibrating equipment, and/or wind on tall structures and process vessels.

For condition a., sleeves should be filled with grout after installation. However, for condition b., sleeves should be sealed on top or filled with an approved elastomeric material to prevent grout or water from filling the sleeve.

4.2.1 Types of Sleeves

Two types of sleeves are commonly used with anchor bolts. The first type is a partial sleeve as shown in Figure 4.1, which is typically used for alignment purposes only. The second type is a full sleeve as shown in Figure 4.2, which is used for alignment purposes as well as for pretensioning the bolt.

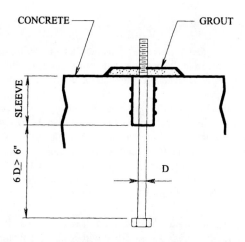

Figure 4.1: Sleeve Used for Alignment Purposes

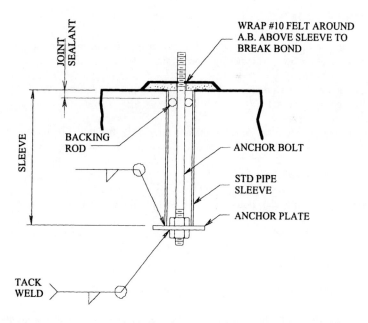

Figure 4.2: Sleeve Used for Alignment Purposes and Pre-Tensioning

4.2.2 Design Considerations

Sleeves will not affect the design of headed anchor bolts subjected to tensile loads, because the tension in the bolt is transferred to the concrete via the anchor bolt head and not by the bond between the bolt and the concrete.

In the design of an anchor bolt with a partial sleeve, the embedment depth of the bolt should be determined as recommended in Chapter 3. However, the distance between the bottom of the sleeve and the anchor-bearing surface should be sufficient to ensure that the concrete below the sleeve will not fail in shear from tensile loads causing the bolt head to snap through the sleeve. The minimum distance between the bottom of the sleeve and the anchor bearing surface should not be less than 6-bolt diameters or 6 inches (150 mm), whichever is greater.

The applied shear force may be resisted by anchor bolts only if the sleeves are filled with grout. If the sleeves are not filled with grout, the anchor bolts will not be effective in resisting the applied shear force. The sleeve, combined with isolation of the bolt from the grout, is desirable to prevent short radius flexing of the anchor bolt due to a horizontal component of the vibration or as a result of thermal growth of the equipment and is an effective way to avoid the most common failure mode of compressor anchor bolts.

4.3 PRETENSIONING

Certain conditions make it desirable to pretension anchor bolts to enhance the performance of the bolt or the performance of the system.

The recommended pretension load is one-third the tensile strength of the bolt unless otherwise required.

Anchor bolts may need to be retightened one week after initial pretensioning to compensate for pre-load losses from strain relaxation within the system.

4.3.1 Pretension Applications

Pretensioning of anchor bolts should be used for the following situations:

a. Tall process towers sensitive to wind (as a rule of thumb these are towers over 100 feet (30 meters) tall or with a height-to-diameter ratio of 15 or more).

b. Reciprocating compressors or other pulsating or vibrating equipment.

c. High-strength anchor bolts (to prevent load reversals on bolts susceptible to fatigue weakening).

4.3.2 Development

Pretensioned anchor bolts should be designed for an embedment development of at least 80% of their ultimate capacity (0.8 f_{ut}).

4.3.3 Methods

Methods that may be utilized to apply the required pre-load are as follows:

a. Hydraulic jacking

Hydraulic jacking is the most accurate method and is recommended if the hydraulic equipment is available and if the physical clearances that exist around each bolt permit its use.

b. Turn-of-nut

Turn-of-nut is the easiest to perform by a construction crew and gives a reasonably accurate result provided that:

* Conditions of grout and base plate can give a consistent "snug-tight" result

* Stretching (spring) length of bolt can be accurately determined.

$$\text{Nut Rotation} = \frac{360 \, l \, A_t \, f_t \, T_{lc}}{E \, A_d} \qquad (4.1)$$

where:

l	=	bolt stretch length, in (mm)
A_t	=	tensile stress area of bolt, in² (mm²)
f_t	=	desired tensile stress, ksi (kPa)
T_{lc}	=	bolt threads per unit length, thds/in (thds/mm)
E	=	elastic modulus of bolt, ksi (kPa)
A_d	=	nominal bolt area, in² (mm²)

If the bolt is to be retightened to compensate for any loss of pre-load, this method requires that nuts be loosened, brought to a "snug tight" condition, and then turned the number of degrees originally specified.

c. Torque wrench

Torque wrench pretensioning provides only a rough measure of actual pretension load but can be the method of choice if equipment for item a. is not available and stretch length of anchor cannot be fixed as required by the "turn of nut" method. The *Industrial Fastener Institute* recommends the following formula for determining the proper tightening torque.

$$T = KDP \qquad (4.2)$$

where:

T = tightening torque, kip-in (kN-mm)
K = torque coefficient, dimensionless
D = fastener diameter, nominal, in (mm)
P = bolt tensile load, kips (kN)

"K" varies from 0.06 to 0.35 (use 0.20 for typical anchor bolt)

4.3.4 Stretching Lengths

Pretensioning should only be implemented when the stretching (spring) length of the anchor bolt extends down to near the anchor head of the bolt. On a typical anchor bolt embedment, as a pre-load is placed upon the bolt, the bolt starts to shed its load to the concrete through its grip (bond) on the bolt. At that time, there exists a high bond stress at the first few inches of embedment. This bond will relieve itself over time and thereby reduce the pre-load on the bolt. Therefore, it is important that the bond be prevented on anchor bolts to be pretensioned. Bond on the bolt shaft can be prevented by wrapping the shaft with plastic tape or by heavily coating the bolt with grease immediately before placing concrete. Grout must not be allowed to bond to the anchor bolt. Tape the portion of the anchor bolt through the grout zone and to within one inch (25 mm) of the bolt head, below the sleeve. (See Figure 4.3)

Tape or grease should not be applied closer than one inch (25 mm) to the anchor bolt head or anchor plate. Anchor bolt sleeves should not be positioned closer than 6D to the bolt head to preclude failure by the head of the bolt pulling through the sleeve.

Sleeved anchor bolts to be pretensioned should have that portion of the bolt beneath the sleeve taped or greased.

The stretching length of the bolt which is pre-loaded within the elastic range acts as a spring in clamping the base plate down against the foundation.

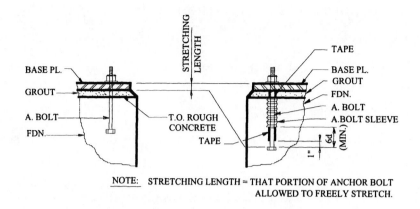

NOTE: STRETCHING LENGTH = THAT PORTION OF ANCHOR BOLT ALLOWED TO FREELY STRETCH.

Figure 4.3: Anchor Bolt Stretch Length

4.3.5 Tightening Sequence

Anchor bolts should be tightened in two stages:

a. First stage should apply 50% of full pre-tension load to all bolts.

b. Second stage should apply full pre-tension load to all bolts.

Bolts should be tightened in a criss-cross pattern. (See Figure 4.4 for circular bolt pattern sequence.)

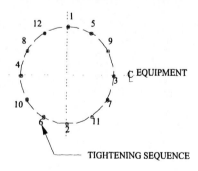

Figure 4.4: Anchor Bolt Tightening Sequence

4.4 CONSIDERATIONS FOR VIBRATORY LOADS

4.4.1 Sleeves

Provide sleeves on all anchor bolts installed on vibratory equipment and isolate bolt from any grout (see Figure 4.5).

4.4.2 Pretensioning

Pretension all anchor bolts installed on vibratory equipment, unless specifically prohibited by the manufacturer.

This pretensioning (stretching) of the anchor bolts creates a spring effect that will absorb the vertical amplitude of the vibration without fatiguing. This spring effect also serves well in clamping the equipment base against the grout without the nut loosening if the amount of anchor bolt stretch exceeds the vertical amplitude of the vibration.

4.5 CONSIDERATIONS FOR SEISMIC LOADS (ZONES 3 AND 4)

Anchorage capacity, including capacity of reinforcement, must exceed minimum specified tensile strength (based on f_{ut}) of the bolt to ensure that any reserve capacity of the bolt can be utilized and that the failure mode will be ductile and in the bolts.

Friction capacity from gravity loads shall not be considered effective in carrying any seismic lateral loads.

Friction capacity may be considered if anchor bolts are pretensioned to twice the calculated seismic uplift force. Friction may then be considered, except friction shall not exceed 50% of that provided by pretension loads.

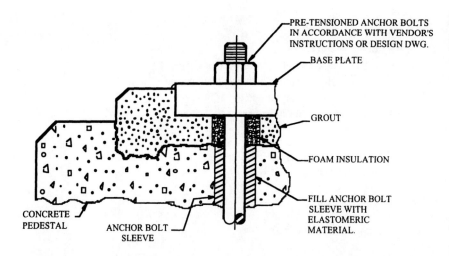

Figure 4.5: Detail for Anchor Bolt for Vibrating Equipment

NOMENCLATURE

A_d = nominal bolt area
A_{eff} = effective anchor bolt area for resisting tension
A_{st} = the area of vertical pier reinforcement per bolt
A_{sv} = area of cross-section of one leg of tie
A_t = anchor bolt tensile stress area

BC = bolt circle diameter
B_n = nominal bearing capacity

D = fastener diameter

E = elastic modulus of bolt

f_t = actual tensile stress
F_t = allowable tensile stress
f_v = actual shear stress
F_v = allowable shear stress
f_y = anchor bolt yield stress
F_y = minimum specified yield strength of reinforcement steel
K = torque coefficient

l = bolt stretch length

M = maximum moment on vessel

n = number of legs in the top 2 sets of ties resisting the shear force (V_u)
N = number of anchor bolts

P = bolt tensile load
P_n = nominal tensile capacity of bolt

T = tightening torque
T_{lc} = bolt threads per unit length
T_u = factored tensile load per bolt

V_u = factored shear force resisted by anchor bolt(s) located in the pier
V_{ua} = factored shear force per bolt

W = minimum weight of vessel

μ = friction coefficient
ϕ = strength reduction factor
ϕ_1 = strength reduction factor for tension load
ϕ_2 = strength reduction factor for shear load

REFERENCES

ACI AB-81, American Concrete Institute, *Guide to the Design of Anchor Bolts and other Steel Embedments.*

ACI 222, American Concrete Institute, *Corrosion of Metal in Concrete.*

ACI 318-89, American Concrete Institute, *Building Code Requirements for Reinforced Concrete.*

ACI 349-82, American Concrete Institute, *Code Requirements for Nuclear Safety Related Concrete Structures.*

ACI 349-90, American Concrete Institute, *Code Requirements for Nuclear Safety Related Concrete Structures.*

ACI 355.1-R91, American Concrete Institute, *State of the Art Report on Anchorage to Concrete.*

AISC ASD, American Institute of Steel Construction, *Specification for Structural Steel Buildings, Allowable Stress Design and Plastic Design,* June 1, 1989.

AISC ASD Manual, American Institute of Steel Construction, *Manual of Steel Construction: Allowable Stress Design,* Ninth Edition, 1989.

AISC LRFD, American Institute of Steel Construction, *Manual of Steel Construction: Load and Resistance Factor Design,* First Edition, 1986.

AISC LRFD Manual, American Institute of Steel Construction, *Load and Resistance Factor Design Specification for Structural Steel Buildings,* September 1, 1986.

API 620, American Petroleum Institute, *Recommended Rules for Design and Construction of Large, Welded, Low-Pressure Storage Tanks,* Seventh Edition, September 1982 (Revision 1-April 1985).

API 650, American Petroleum Institute, *Welded Steel Tanks for Oil Storage,* Ninth Edition, July 1993.

ASTM A36, American Society for Testing and Materials, *Specification for Structural Steel,* 1991.

ASTM A143, American Society for Testing and Materials, *Practice for Safeguarding Against Embrittlement of Hot-Dip Galvanized Structural Steel Products and Procedure for Detecting Embrittlement,* 1974 (Revised 1989).

ASTM A153, American Society for Testing and Materials, *Specification for Zinc Coating (Hot-Dip) on Iron and Steel Hardware,* 1987.

ASTM A193, American Society for Testing and Materials, *Specification for Alloy-Steel and Stainless Steel Bolting Materials for High-Temperature Service,* 1992.

ASTM A307, American Society for Testing and Materials, *Specification for Carbon Steel Bolts and Studs,* 60,000 psi Tensile, 1992.

ASTM A325, American Society for Testing and Materials, *Specification for Structural Bolts, Steel, Heat-Treated, 120/105 ksi Minimum Tensile Strength,* 1992.

ASTM A449, American Society for Testing and Materials, *Specification for Quenched and Tempered Steel Bolts and Studs,* 1992.

ASTM A490, American Society for Testing and Materials, *Specification for Heat-Treated, Steel Structural Bolts, 150 ksi (1035 MPa) Tensile Strength,* 1992.

ASTM A588, American Society for Testing and Materials, *Specification for High-Strength, Low-Alloy Structural Steel with 50 ksi (345 MPa) Minimum Yield Point to 4 in. (100 mm) Thick,* 1991.

ASTM A767, American Society for Testing and Materials, *Specification for Zinc-Coated (Galvanized) Bars for Concrete Reinforcement,* 1990.

ASTM A775, American Society for Testing and Materials, *Specification for Epoxy-Coated Reinforcing Steel Bars,* 1992.

ASTM A780, American Society for Testing and Materials, *Practice for Repair of Damaged and Uncoated Areas of Hot-dip Galvanized Coatings,* 1992.

Bailey and Burdette, John W. Bailey and Edwin G. Burdette, *Edge Effects on Anchorage to Concrete,* Civil Engineering Research Series No. 31, The University of Tennessee at Knoxville, 1977.

Blodgett, Omer Blodgett, *Design of Welded Structures,* The James F. Lincoln Arc Welding Foundation, Cleveland, Ohio, 1966.

Cook and Klingner, Ronald A. Cook and Richard E. Klingner, "Behavior of Ductile Multiple Anchor Steel to Concrete Connections with Surface Mounted Baseplates", ACI SP130, *Anchors in Concrete Design and Behavior,* 1991.

Fuchs et al, Werner Fuchs, Rolf Eligehausen, John Breen, *Concrete Capacity Design (CCD) Approach for Fastening to Concrete*, ACI Structural Journal, Vol. 92, No. 1, January/February, 1995.

Furche et al, Johannes Furche, and Rolf Eligehausen, "Lateral Blowout Failure of Headed Studs near a Free Edge", ACI SP130 *Anchors in Concrete -- Design and Behavior"*, American Concrete Institute, Detroit, 1991.

Hasselwander, Jirsa, Breen and Lo, G.B. Hasselwander, J.O. Jirsa, J.E. Breen, and K. Lo, *Strength and Behavior of Anchor Bolts Embedded Near Edges of Concrete Piers,* Research Report 29-2F, Center for Highway Research, The University of Texas at Austin, 1977.

Industrial Fasteners Institute, Fastener Standards, Sixth Edition, p. M-64.

Lee and Breen, D.W. Lee and J.E. Breen, *Factors Affecting Anchor Bolt Development,* Research Report 88-1F, Center for Highway Research, The University of Texas at Austin, 1966.

Steel Design Guide Series, Volume 1, John T. Dewolf, *Steel Design Guide Series, Design of Column Base Plates*, American Institute of Steel Construction, 1990.

Steel Design Guide Series, Volume 7, James M. Fisher, *Steel Design Guide Series, Industrial Buildings, Roofs to Column Anchorage*, American Institute of Steel Construction, 1993.

Uniform Building Code, International Conference of Building Officials, Whittier, California, 1991.

INDEX

A

Allowable stress design 3-13
American Concrete Institute (ACI) 3-1, 3-5, 3-15, 3-18, 3-21; ACI 349, Appendix B 1-1, 1-2, 3-1—3-3, 3-14; ACI Publication AB-81 1-1; and CCD 3-1; committees 1-2; corrosion codes and specifications 2-3—2-4
American Institute of Steel Construction (AISC) 2-4, 3-13
American Petroleum Institute (API) 2-4
American Society for Testing Materials (ASTM) 2-5, 2-6, 3-3, 3-18; and bolt materials 2-2
Anchorage capacity 3-1—3-3
Anchorage failure 3-6, 3-10—3-16
Anchor bolts (cast-in-place, headed bolts): coatings and corrosion protections 2-1—2-7; codes and specifications 1-1—1-2, 2-3—2-4; common materials 2-2; configuration 3-4; design considerations 3-3—3-7; design methods 3-1—3-3; failure modes 3-10—3-16; force distribution 3-7—3-10; pretension applications 4-3—4-6; reinforcement systems 3-16—3-22; sleeves 4-1—4-3; tightening sequence 4-6; vibratory and seismic loads 4-7—4-8

B

Bearing failure, localized 3-14—3-16
Biaxial loads, design method changes 3-2
Bolt configuration 3-3—3-4
Bolt loads 3-11
Bolt shear 3-21—3-22

Bolt tension 3-17—3-19

C

Coatings, corrosion prevention 2-4—2-7
Codes 1-2, 2-3—2-6
Cold-applied zinc 2-6
Concrete bearing strength 3-15—3-16
Concrete Capacity Design (CCD) method 1-2, 3-1—3-3
Concrete splitting failure 3-16
Cone model 3-1—3-2, 3-14
Configurations: anchor bolts (cast-in-place, headed bolt) 3-4; design flowchart 3-6; octagonal pedestal reinforcement 3-20; square and rectangular pedestals reinforcement 3-19; stretch length 4-6; tightening sequence 4-6
Corrosion 2-1, 2-3—2-7
Corrosion allowance 2-5
Corrosion rates 2-4

D

Design basis 3-4—3-7
Design load, considerations 3-5
Design methods 1-2, 3-1—3-3
Ductile connections: defined 3-5, 3-7; and shear 3-10

E

Electro-deposited zinc coating 2-6
Environmental conditions, corrosion 2-3
Equations: allowable stress design expression (3.6) 3-13; bearing strength, B_n (3.7) 3-14—3-16; concrete pullout and fastener strengths

(3.8-3.10) 3-15—3-16; maximum tension (3.1) 3-8; nut rotation (4.1) 4-4; reduced anchor bolt area, A_{eff} (3.2) 3-9; shear, reinforcement area, A_{sv} (3.12) 3-21; torque formula, T (4.2) 4-5; ultimate strength design, P_n (3.3-3.5) 3-12—3-13; vertical pier reinforcement area, A_{st} (3.11) 3-17—3-18

F
Factored service loads 3-5
Failure modes 3-10—3-16
Fastener strength 3-15—3-16
Fireproofing 2-6
Force distribution 3-7—3-10
Foundation designs 3-3
Full sleeve 4-1, 4-2

G
Galvanized bolts 2-1, 2-5—2-6
Grades, anchor bolt materials 2-1, 2-2

H
Hot dip zinc 2-5—2-6
Hydraulic jacking method 4-4

I
Industrial Fastener Institute 4-5
Insulation 2-6

L
Lateral bursting failure 3-14
Loads, bolt 3-11

M
Moment of inertia 3-8

N
Nonductile connections: defined 3-7; and shear 3-10

O
One-third Ld space requirement 3-18

P
Partial sleeve 4-1, 4-2, 4-3
Pickling process 2-1
Pier design 3-16—3-17
Pier reinforcement 3-17—3-22
Plates, bearing 3-14—3-16
Pretension applications 4-3—4-7
Pullout failure 3-13—3-14
Pyramid model 3-1—3-2, 3-14

S
Seismic loads 4-7
Shear: design method changes 3-2; distribution of 3-10; forces 3-21—3-22; shear cone 3-13—3-14; and tension 3-12—3-13
Sleeves 4-1—4-3
Specifications, and codes 2-3—2-6
Splitting failure, concrete 3-16
Strength design 3-12—3-13
Stress design, bolt failure 3-13
Stretching (spring) lengths 4-5—4-6
Structural column anchorage 3-8—3-10

T
Tensile stress: bolt failure 3-11—3-14; lateral bursting (blowout) failure 3-18
Tension: design method changes 3-2; equation 3-8; lateral bursting force 3-18; and shear 3-12—3-13; vertical pier reinforcement 3-17—3-18
Torque wrench method 4-4—4-5
Turn-of-nut method 4-4

U
Ultimate strength design 3-12—3-13
Ultimate tensile capacity 3-5
Uniform Building Code (UBC) 3-12

V
Vertical pier reinforcement area 3-17—3-18
Vertical vessel anchorage 3-8
Vibratory loads 4-7

W
Weathering steels 2-6—2-7